T0305819

Dendritic Cell Interactions with Bacteria

Emerging evidence suggests that dendritic cells play a major role in the orchestration of the immune response to bacteria. This book introduces the reader to the complex world of dendritic cells and describes how the intimate interplay between dendritic cells, bacteria and the environment dictates either the induction of immunity or tolerance to the encountered micro-organisms. It discusses how this can allow organisms to tolerate beneficial bacteria and to react against pathogens, as well as the strategies pathogenic bacteria have evolved to escape dendritic cell patrolling. Expert contributors discuss everything from bacterial capture and recognition to their killing, processing and the induction of adaptive immunity. Particular focus is on the tissue context in which bacteria are handled by dendritic cells and on possible defects therein, which may potentially lead to chronic infection or inflammation. Graduate students and researchers will find this an invaluable overview of current dendritic cell biology research.

MARIA RESCIGNO is the Director of the Immunobiology of Dendritic Cells and Immunotherapy Research Unit at the European Institute of Oncology, Milan, Italy.

Published titles

Over the past decade, the rapid development of an array of techniques in the fields of cellular and molecular biology has transformed whole areas of research across the biological sciences. Microbiology has perhaps been influenced most of all. Our understanding of microbial diversity and evolutionary biology, and of how pathogenic bacteria and viruses interact with their animal and plant hosts at the molecular level, for example, have been revolutionized. Perhaps the most exciting recent advance in microbiology, a fusion of classical microbiology, microbial molecular biology and eukaryotic cellular microbiology. Cellular microbiology is revealing how pathogenic bacteria interact with host cells in what is turning out to be a complex evolutionary battle of competing gene products. Molecular and cellular biology are no longer discrete subject areas but vital tools and an integrated part of current microbiological research. As part of this revolution in molecular biology, the genomes of a growing number of pathogenic and model bacteria have been fully sequenced, with immense implications for our future understanding of microorganisms at the molecular level.

Advances in Molecular and Cellular Microbiology is a series edited by researchers active in these exciting and rapidly expanding fields. Each volume will focus on a particular aspect of cellular or molecular microbiology and will provide an overview of the area, as well as examine current research. This series will enable graduate students and researchers to keep up with the rapidly diversifying literature in current microbiological research.

Series Editors

Professor Brian Henderson
University College, London

Professor Michael Wilson
University College, London

Professor Sir Anthony Coates
St George's Hospital Medical School, London

Professor Michael Curtis
St Bartholomew's and Royal London Hospital, London

Advances in Molecular and Cellular Microbiology 14

Dendritic Cell Interactions with Bacteria

EDITED BY
MARIA RESCIGNO
European Institute of Oncology

CAMBRIDGE
UNIVERSITY PRESS

CAMBRIDGE
UNIVERSITY PRESS

Shaftesbury Road, Cambridge CB2 8EA, United Kingdom

One Liberty Plaza, 20th Floor, New York, NY 10006, USA

477 Williamstown Road, Port Melbourne, VIC 3207, Australia

314–321, 3rd Floor, Plot 3, Splendor Forum, Jasola District Centre, New Delhi – 110025, India

103 Penang Road, #05–06/07, Visioncrest Commercial, Singapore 238467

Cambridge University Press is part of Cambridge University Press & Assessment, a department of the University of Cambridge.

We share the University's mission to contribute to society through the pursuit of education, learning and research at the highest international levels of excellence.

www.cambridge.org
Information on this title: www.cambridge.org/9780521855860

First published 2007

A catalogue record for this publication is available from the British Library

Library of Congress Cataloging-in-Publication data
Dendritic cell interactions with bacteria/edited by Maria Rescigno.
    p. ; cm. – (Advances in molecular and cellular microbiology; 14)
Includes bibliographical references and index.
ISBN-13:   978-0-521-85586-0 (hardback)
1. Dendritic cells. 2. Bacteria. 3. Host-bacteria relationships. 4. Bacterial
    diseases–Immunological aspects. I. Rescigno, Maria, 1968-II. Title. III. Series.
[DNLM: 1. Dendritic Cells–immunology. 2. Bacteria–immunology.
3. Bacterial Infections–immunology. QW 568 D3905 2007]
QR185.8.D45D452 2007
616.07'9–dc22   2006020611

ISBN   978-0-521-85586-0   Hardback

Contents

CONTENTS

Preface

Dendritic cells (DCs) comprise a family of professional antigen presenting cells that are unique in their ability to activate T lymphocytes. Dendritic cells patrol all the tissues at the interface with the external world, including skin and mucosal surfaces, for the presence of invaders. The DC system is characterized by a remarkable plasticity that allows the induction both of immunity and tolerance toward the encountered antigens. This is achieved through the combination of a number of different factors, including the subsets of DCs, their activation state and environmental cells that can regulate DC function. DCs are present in the periphery in an immature form that is particularly apt at capturing antigens and at deciphering the messages associated therein. After an activation stimulus that is delivered by some antigens (including bacteria) or by inflammatory cytokines released during inflammation, activated DCs acquire a migratory phenotype and reach the draining lymph node. Here, DCs present the antigens captured in the periphery and initiate T cell adaptive immune responses.

This book describes how the intimate interplay between dendritic cells, bacteria and the environment dictates the induction of immunity or tolerance to bacteria and how pathogenic bacteria have evolved strategies to escape DC patrolling. The first section introduces the complexity of the DC system describing the different subpopulations of DCs and their role in the induction of immune responses. This is followed by the description of a class of pathogen recognition receptors and their signaling pathways that are fundamental in the activation of DCs after recognition of bacterial structural components. These receptors, belonging to the Toll-like receptor family, are differentially expressed on DC subpopulations and contribute to generate functional diversity. To conclude this general part on DC function, there is a description on how bacterial antigens are handled, processed and presented by DCs.

In the second section, attention switches to the role of DCs in the initiation and orchestration of innate immune responses. The section begins describing how dendritic cells can directly participate in the uptake of bacteria across mucosal surfaces and its consequences in terms of DC activation. After microbial recognition, DCs act first as innate immune cells that release inflammatory mediators that can strengthen and amplify the innate immune response. In particular a novel monocyte-derived DC population called TipDCs that produces large amounts of tumor necrosis factor (TNF) and inducible nitric oxide synthase (iNOS) is reported. Then DCs can leave the infected site to reach the draining lymph node for T cell activation. Thus, DCs represent a link between innate and adaptive immunity because their activation can lead on one side to the recruitment and activation of innate immune cells like granulocytes, macrophages and natural killer (NK) cells and on the other side to the activation of adaptive immune cells. To achieve this, DCs can act on their own or in concert with other innate immune cells like NK cells, as discussed in the last chapter of this section.

The following section deals with the initiation of adaptive immune responses that is conducted by DCs that have deciphered and integrated signals deriving from the bacteria, the infected tissue and the recruited immune cells. Two major examples of DC handling of strictly or facultative intracellular bacteria have been considered, namely *Legionella* and *Salmonella*. It is described how differently from macrophages, DCs have evolved strategies to handle and control intracellular growth of *Legionella* and to activate effective adaptive immune responses to control bacterial infection. Interestingly, DCs can present bacterial antigens also when they are non-infected after phagocytozing infected cells. This process also known as cross-presentation is unique to DCs and favors the activation of T cell responses toward *Salmonella, Listeria* and *Mycobacterium.*

Finally, strategies developed by bacteria to evade DC recognition and activation are discussed in the fourth section. Here pathogen recognition receptors are thoroughly discussed as possible targets for pathogens to modulate immune function of antigen presenting cells. It is described that the cross-talk between different classes of pathogen recognition receptors can lead to suppression or activation of immune responses. In the following chapter the ability of bacteria or their products to suppress the immune response through the skewing of T cell responses toward regulatory T cells or to subtypes which are inappropriate for bacterial elimination is reported.

A major drawback of improper bacterial handling can result in chronic inflammatory responses particularly at sites continuously exposed to bacteria like the gut. Here, commensal bacteria are beneficial to the host as they help digesting ingested food through the degradation of complex sugars and metabolites. In order to tolerate ''good'' bacteria, the immune system has developed strategies to cohabitate with beneficial bacteria and discriminate harmful pathogens. When these strategies are disrupted, inflammatory responses can arise leading to inflammatory bowel disease as discussed in the last chapter of this section.

In conclusion, this book has brought together experts in several fields of dendritic cell—bacteria interaction from their capture and recognition to their killing, processing and induction of adaptive immunity. Much attention has been focused on the tissue context where bacteria are handled by DCs. When defects either in bacterial handling or in the interaction with the environment are encountered, chronic infection or inflammation can arise.

Abbreviations

APC	antigen-presenting cell
ASK	apoptosis signal-regulating kinase
BCG	bacillus Calmette-Guerin
BIR	baculoviral inhibitors of apoptosis repeat
CARD	caspase recruitment domain
CD	Crohn's disease
cDC	conventional DC
CLP	common lymphoid progenitor
CLR	C-type lectin-related
CMP	common myeloid progenitor
CRD	carbohydrate-recognition domain
CT	cholera toxin
CTL	cytotoxic T lymphocytes
DALIS	dendritic cells aggresome-like induced structures
DC	dendritic cell
DRIP	defective ribosomal product
dsRNA	double-stranded RNA
DSS	dextran sodium sulfate
EC	epithelial cell
ER	endoplasmic reticulum
ERAD	ER-associated degradation
ERAP	endoplasmic reticulum aminopeptidase
FADD	Fas (TNFRSF6)-associated via death domain
FAE	follicle-associated epithelium
GALT	gut associated lymphoid tissue
GFP	green fluorescent protein
GM-CSF	granulocyte-macrophage colony-stimulating factor

HCV	Hepatitis C virus
HLA	human leukocyte antigen
IAP	inhibitors of apoptosis
IBD	inflammatory bowel disease
IDC	immature DC
IE-DAP	γ-δ-glutyl-meso diaminopimelic acid
IFN	interferon
Ii	invariant chain
IKK	IκB kinase
IL	interleukin
iNOS	inducible nitric oxide synthase
IRAK	IL-1R-associated kinase
IRF	interferon regulatory factor
ISGF	IFN-stimulated gene factor
ISRE	IFN-stimulated regulatory element
ITAM	immunoreceptor tyrosine-based activation motif
JNK	c-Jun N-terminal kinase
KIR	killer Ig-like receptors
LAM	lipoarabinomannan
LLO	listeriolysin O
LP	lamina propria
LPS	lipopolysaccharide
LRR	leucine-rich repeat
LTA	lipoteichoic acid
mAB	monoclonal antibody
MAL	MyD88 adaptor-like
MAPKK	mitogen activated protein kinase kinase
MAPKKK	mitogen activated protein kinase kinase kinase
MDP	muramyl dipeptide
MEF	mouse embryonic fibroblast
MHC	major histocompatibility complex
MLN	mesenteric lymph nodes
NCR	nitrogen catabolite repressor
NDV	Newcastle disease virus
NEMO	NF-κB essential modulator
NF	nuclear factor
NK	natural killer
NOD	nucleotide-binding oligomerization domain
Nod-LRR	nucleotide oligomerization domain-leucine-rich repeat
OVA	chicken ovalbumin

PAMP	pathogen associated molecular patterns
pDC	plasmacytoid DC
PGN	peptidoglycan
PI3P	phosphoinositol-3-phosphate
PKR	protein kinase R
PP	Peyer's patches
PPAR	peroxisome-proliferator-activated receptor
PRR	pathogen recognition receptor
RICK	Rip-like interacting caspase-like apoptosis-regulatory protein kinase
RIG	retinoic acid-inducible protein
RIP	receptor interacting protein
SARM	sterile α and HEAT-Armadillo motif
siRNA	small interfering RNA
SLE	systemic lupus erythematosus
SPI	*Salmonella* pathogenicity island
ssRNA	single-stranded RNA
STAT	signal transducer and activator of transcription
TAB	tubulin antisense-binding protein
TAK	TGFβ-activating kinase
TAP	transporter associated with antigen processing
TBK	TANK-binding kinase
TGF	transforming growth factor
TipDC	tumor infiltrating pDC
TIR	Toll/IL1 receptor
TIRAP	TIR domain-containing adaptor protein
TJ	tight junction
TLR	Toll-like receptor
TNF	tumor necrosis factor
TRAM	TRIF-related adaptor molecule
TRIF	TIR domain-containing adaptor inducing IFNβ
TSLP	thymic stromal lymphopoietin
VSV	Vesicular stomatis virus

Contributors

Shizuo Akira
Department of Host Defense
Research Institute for Microbial
    Diseases
Osaka University
3-1 Yamada-oka
Suita
Osaka 565-0871
Japan
    and
ERATO, Japan Science and
    Technology Agency
3-1 Yamada-oka
Suita
Osaka 565-0871
Japan

Carlos Ardavín
Department of Immunology and
    Oncology
Centro Nacional de Biotecnologia/
    CSIC
Universidad Autònoma
28049 Madrid
Spain

Christoph Becker
I Department of Medicine
University of Mainz
55131 Mainz
Germany

Laurence Bougnères-Vermont
Institut Curie
Inserm u653
26 rue d'Ulm
75248 Paris
cedex 05
France

Miriam T. Brady
Immune Regulation Research
    Group
School of Biochemistry and
    Immunology
Trinity College
Dublin 2
Ireland

Anneke Engering
Department of Molecular
    Cell Biology and
    Immunology
VU Medical Center
v.d. Boechorststraat 7
1081 BT Amsterdam
The Netherlands

Guido Ferlazzo
Istituto Nazionale Ricerca sul
    Cancro
Genoa
Italy
    and
University of Messina
Messina 98100
Italy

Teunis B. H. Geijtenbeek
Department of Molecular Cell
    Biology and Immunology
VU Medical Center
v.d. Boechorstraat 7
1081 BT Amsterdam
The Netherlands

Pierre Guermonprez
Institut Curie, inserm u653
26 rue d'Ulm
75248 Paris
cedex 05
France

Estella A. Koppel
Department of Molecular Cell
    Biology and Immunology
VU Medical Center
v.d. Boechorstraat 7
1081 BT Amsterdam
The Netherlands

Peter McGuirk
Opsona Therapeutics
Biotechnology Building
Trinity College
Dublin
Ireland

Kingston H. G. Mills
Immune Regulation Research
    Group
School of Biochemistry and
    Immunology
Trinity College
Dublin 5
Ireland

Catarina Nogueria
Section of Microbial Pathogenesis
Yale University School of Medicine
Boyer Center for Molecular Medicine
295 Congress Avenue
New Haven
CT 06536

Eric G. Pamer
Infectious Diseases Service
Memorial Sloan-Kettering Cancer
    Center
Immunology Program
Sloan Kettering Institute
1275 York Avenue
New York
New York 10021
USA

Maria Rescigno
European Institute of Oncology
Department of Experimental
    Oncology
Via Ripamonti 435
20141 Milan
Italy

Craig R. Roy
Section of Microbial Pathogenesis
Yale University School of Medicine
Boyer Center for Molecular
    Medicine
295 Congress Avenue
New Haven
CT 06536
USA

Natalya V. Serbina
Infectious Diseases Service
Memorial Sloan-Kettering Cancer
    Center
Immunology Program
Sloan Kettering Institute
1275 York Avenue
New York
New York 10021
USA

Sunny Shin
Section of Microbial Pathogenesis
Yale University School of Medicine
Boyer Center for Molecular
    Medicine
295 Congress Avenue
New Haven
CT 06536
USA

Osamu Takeuchi
Department of Host Defense
Research Institute for Microbial
    Diseases
Osaka University
    and
ERATO,
Japan Science and Technology
    Agency

3-1 Yamada-oka
Suita
Osaka 565-0871
Japan

Yvette van Kooyk
Department of Molecular Cell
    Biology and Immunology
VU Medical Center
v.d. Boechorststraat 7
1081 BT Amsterdam
The Netherlands

Sandra J. van Vliet
Department of Molecular Cell
    Biology and Immunology
VU Medical Center
v.d. Boechorststraat 7
1081 BT Amsterdam
The Netherlands

Mary Jo Wick
Department of Microbiology and
    Immunology
Göteborg University
Box 435
SE 40530 Göteborg
Sweden

xix

CONTRIBUTORS

PART I Dendritic cells and their role in immunity

CHAPTER 1

Subpopulations and differentiation of mouse dendritic cells

Carlos Ardavín
Universidad Autónoma

1.1 DENDRITIC CELL SUBPOPULATIONS

Dendritic cells (DCs) have an essential function in the immune system by participating in primitive defense responses that constitute the innate immunity, as well as in the induction and regulation of antigen-specific immune responses. This allows DCs to control infections caused by parasitic and microbial pathogens, to block tumour growth and to exert a precise regulation of T cell, B cell and NK cell immune responses. In addition, DCs also fulfill a pivotal role in the induction and maintenance of T cell tolerance. The functional diversity characterizing the DC system relies essentially on the remarkable plasticity of the DC differentiation process, which dictates the acquisition of DC functional specialization through the generation of a large collection of DC subpopulations (reviewed by Shortman and Liu, 2002). Dendritic cells are located both in the lymphoid organs (such as the spleen or the lymph nodes), and in non-lymphoid tissues (such as the skin or the liver), and can be classified in two major categories: conventional DCs (cDCs), and plasmacytoid DCs (pDCs). Whereas in turn cDCs comprise multiple DC subpopulations endowed with specific functions, little is known about the functional heterogeneity of pDCs. A summary of the most relevant phenotypic and functional characteristics of the main DC subpopulations present in mice is shown in Table 1.1.

A first group of cDCs includes those that are common, and largely restricted, to the majority of organized lymphoid organs of the immune system, and perform their specific functions, as immature or mature DCs, within these organs. This group of lymphoid organ-restricted cDCs comprises three main DC subpopulations, namely $CD8^+$ $CD11b^-$ DCs (herein called $CD8^+$ DCs), $CD8^-$ $CD11b^+$ DCs (herein called $CD8^-$ DCs), and $CD8^-$ $CD11b^-$ DCs (herein called $CD11b^-$ DCs). $CD8^-$ DCs

Table 1.1. *Location and phenotypic characteristics of the principal mouse DC subpopulations*

Location	pDCs	CD8$^+$ DCs	CD8$^-$ CD4$^+$ DCs	CD8$^-$ CD4$^-$ DCs	CD11b$^-$ DCs	Langerhans cells	Dermal DCs
Location							
Thymus	Yes	Yes	No	Yes[a]	Few	No	No
Spleen	Yes	Yes	Yes	Yes	No	No	No
Skin	[b]	No	No	No	No	Yes	Yes
Peripheral LNs	Yes	Yes	Few	Yes	Few	Yes	Yes
Intestinal Peyer's patches	Yes	Yes	Few	Yes	Yes	Yes	No
Mesenteric LNs	Yes	Yes	Few	Yes	Yes	No	[c]
Other locations	Bone marrow, blood, lung	Liver		Liver	Liver	Epidermis, intestinal S-E, lung S-E, vagina S-E	Intestinal L-P,[d] bronchial L-P,[d] lung interstitium,[d] liver parenchyma,[d] kidney parenchyma,[d] CNS parenchyma[d]

Phenotype

Phenotype							
CD11c	+ int	+ high	+ high	+ high	+ high	+ high	+ high
CD11b	−	+ low	+ high	+ high	+ high	+ high	+ high
CD8	$+^e$	+	−	−	−	− (→ +)^f	−
CD4	$+^e$	−	+	−	−	−	−
B220	+	−	−	−	−	−	−
CD62L	+	−	−	Not analyzed	−	−	−
DEC-205	−	+ high	−	−	+ high	+ high	+ low

Abbreviations: CNS, central nervous system; LNs, lymph nodes; L-P, lamina propria; S-E, stratified epithelium.

Notes

[a] Thymic CD8− DCs have been reported to express low levels of CD4, although it has been suggested that in fact CD8− thymic DCs are CD4−, since CD4 appears to be picked up by thymic DCs from surrounding CD4+ T cells.

[b] pDCs are absent from the skin in steady state, but can be recruited to this location during inflammation.

[c] DCs with similar characteristics than those defining dermal DCs after migration to the peripheral LNs are found in the mesenteric LNs; these cells have been claimed to correspond to intestinal L-P DCs.

[d] Interstitial DCs that have been suggested to be functionally related to dermal DCs have been described in these locations.

[e] pDCs have been demonstrated to upregulate both CD8 and CD4 during activation.

[f] Langerhans cells have been demonstrated to upregulate CD8 during in vivo maturation and migration from the skin to the draining lymph nodes.

can be in turn subdivided in CD8$^-$ CD4$^+$ DCs and CD8$^-$ CD4$^-$ DCs. Whereas CD8$^+$ DCs and/or CD8$^-$ DCs are present in the thymus, spleen, lymph nodes and lymphoid tissue of the intestinal and respiratory tracts (reviewed by Shortman and Liu, 2002), the CD11b$^-$ DC subpopulation appears to be predominantly related to the intestinal lymphoid system (reviewed by Johansson and Kelsall, 2005), representing approximately one-third of the DCs found in the Peyer's patches and mesenteric lymph nodes. CD8$^+$ DCs and CD8$^-$ DCs do not appear to migrate or recirculate to other effector organs of the immune system to fulfill their functions. In this sense, CD11b$^-$ DCs have been tentatively included in this category since they appear to be present in the majority of lymphoid organs (reviewed by Johansson and Kelsall, 2005), but their immunobiology, and particularly their migratory behavior, remains largely unknown.

A second group of cDCs comprises those located, in an immature state, in antigen-uptake sites within non-lymphoid organs. Upon contact with an antigen, these DCs migrate through the lymph vessels to the lymph nodes, where they interact with antigen-specific T cells, and in some cases with other effector cells of the immune system, including DCs, NK cells and B cells. This group of migrating-cDCs comprises Langerhans cells (located in the epidermis, and other stratified and pseudo-stratified epithelia of the intestinal, respiratory and reproductive tracts), and interstitial DCs. These in turn include dermal DCs, and other interstitial DC subpopulations, as those found in the lamina propria of the intestinal and respiratory tracts, as well as those located in the lung interstitium, and in the parenchyma of the liver, kidney and CNS.

A number of experimental evidences suggest that in the steady state both lymphoid-organ restricted cDCs and migrating-cDCs are generated locally from blood-borne, immediate DC precursors, originating in the bone marrow (reviewed by Ardavín, 2003). Interestingly, during ongoing immune responses, a strong increase in the absolute number of both lymphoid organ-restricted cDCs and migrating-cDCs has been reported, raising the problem of the identity of the immediate DC precursors responsible for the generation of these de novo-formed DCs. It could be hypothesized that the same precursors are responsible for the generation of cDCs under steady state and infection. Alternatively, during infection, apart from, or instead of the precursors functioning in steady state, additional inflammatory precursors, with equivalent or complementary DC differentiation potential, could be recruited. Although this remains an open issue in DC biology, different research groups have reported that during inflammation and/or infection, monocytes, which represent so far the best known DC precursors, could be recruited to both peripheral tissues (such as the dermis) and to lymphoid organs, and differentiate locally into DCs (reviewed by Leon *et al.*, 2005). The relationships between monocyte-derived DCs (or other putative de novo-formed inflammatory DCs) and the pre-existing lymphoid organ-restricted cDCs or migrating-cDC subpopulations, has to be addressed. In this sense, data from our laboratory suggest that during infection with *Leishmania major*, monocytes are actively recruited to the popliteal lymph nodes draining the site of infection, where they differentiate into CD11b$^+$ CD8$^-$ DCs, with phenotypic characteristics that differ from those defining the DC subpopulations pre-existing in the popliteal lymph nodes (C. Ardavín *et al.*, unpublished data). Although the role played by monocyte-derived DCs during *Leishmania* infection has to be defined, our results support the concept that the immune responses against certain pathogens could involve the participation of newly-formed DCs endowed with specific functions.

In contrast to cDCs, pDCs appear to differentiate in the bone marrow, and migrate to the blood and lymphoid organs, where they are activated and participate in anti-viral immune responses (reviewed by Colonna *et al.*, 2004). In addition, it has been shown that pDCs can be recruited to the inflamed skin during atopic dermatitis, contact dermatitis or psoriasis, and to systemic lupus erythematosis (SLE)-associated skin lesions (reviewed by Valladeau and Saeland, 2005). However, the possibility that during inflammation and/or infection pDC precursors are recruited to the lymphoid organs and differentiate locally, has not been addressed in depth and therefore cannot be excluded.

An integrated model of the development and function of the mouse DC system is represented in Figure 1.1.

1.1.1 Lymphoid organ-restricted cDCs

In the steady state, CD8$^+$ DCs, CD8$^-$ DCs and CD11b$^-$ DCs are found in the lymphoid organs as immature DCs; however, under situations of inflammation and/or infection DCs enter into a maturation program as a result of the engagement of DC activating receptors (reviewed in Johansson and Kelsall, 2005 and Villadangos and Heath, 2005). Globally, these DC subpopulations are considered to induce T cell responses after capturing antigens that gain access to the lymphoid organs through the blood or lymph vessels. In addition, as discussed below, at least CD8$^+$ DCs can also cross-present antigens captured in the vicinity of epithelial surfaces by migrating DCs that

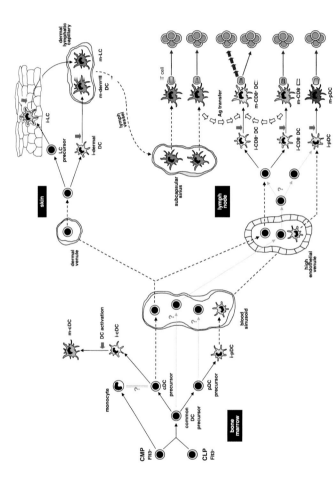

Figure 1.1. Integrated model of the development and function of the mouse DC system. Black solid arrows represent differentiation and/or maturation processes; black broken arrows represent migration processes; grey broken arrows represent hypothetical differentiation or migration processes. Black thick arrows indicate the induction of DC activation initiated by pathogens or inflammatory mediators. White thick arrows indicate antigen transfer. Double arrows indicate T cell direct priming; grey thick arrows indicate T cell crosspriming. Abbreviations: i-DC, immature DC; LC, Langerhans cell; m-DC, mature DC. (For a colour version of this figure, please refer to the colour insert between pages 12 and 13.)

transport these antigens to the lymph nodes and transfer them to CD8$^+$ DCs (Belz *et al.*, 2004).

Maturation of lymphoid organ-restricted DCs could involve a local migration process, from the antigen capture zones to the T cell areas within the same lymphoid organ, but not the migration between different organs through the lymphatics. In particular, it has been reported that upon activation, CD8$^-$ DCs migrate from the outer to the inner part of the splenic white pulp (De Smedt *et al.*, 1996; Reis e Sousa *et al.*, 1997), and from the subepithelial dome to the interfollicular region of the Peyer's patches (Iwasaki and Kelsall, 2000). In general, antigen presentation by immature CD8$^+$ DCs or CD8$^-$ DCs leads to T cell tolerance, whereas antigen uptake by these DC subpopulations in the presence of maturation stimuli leads to the induction of T cell immunity by mature CD8$^+$ DCs or CD8$^-$ DCs (reviewed by Steinman and Nussenzweig, 2002).

CD8$^+$ DCs

In steady state, CD8$^+$ DCs are located in the cortico-medullary zone and medulla of the thymus (reviewed by Ardavín, 1997) (where they participate in the negative selection of the developing thymocytes) and in the T cell areas of the lymphoid organs (reviewed by Villadangos and Heath, 2005) (i.e. inner part of the splenic white pulp, deep cortex of the lymph nodes, and interfollicular regions of the lymphoid aggregates of the intestinal tract). CD8$^+$ DCs are less efficient than CD8$^-$ DCs in capturing antigens by endocytosis or phagocytosis (Leenen *et al.*, 1998). However, they display a high capacity to internalize apoptotic cells (Iyoda *et al.*, 2002), which has been correlated with their ability to induce peripheral tolerance to self tissue-associated antigens in the steady state (Belz *et al.*, 2002; Liu *et al.*, 2002). CD8$^+$ DCs can stimulate CD4$^+$ or CD8$^+$ T cell responses by direct- or cross-priming, after antigen uptake in the lymphoid organs (reviewed by Villadangos and Heath, 2005). Alternatively, it has been proposed that after antigen uptake in non-lymphoid organs by dermal or interstitial DCs, which subsequently migrate to the lymph nodes, these antigens can be cross-presented by lymph node CD8$^+$ DCs, as the result of antigen transfer from the migrating-DCs to resident CD8$^+$ DCs (reviewed by Villadangos and Heath, 2005). In this regard, results from various research groups agree on the idea that CD8$^+$ DCs are the most effective, if not the sole, DC subpopulation capable of stimulating a T cell response by cross-presentation in vivo (den Haan *et al.*, 2000; Pooley *et al.*, 2001), although it has been recently reported that Langerhans cells can stimulate OVA-specific T cells in vivo by cross-priming (Mayerova *et al.*, 2004). The cross-priming capacity of CD8$^+$ DCs has been proposed to rely on their potential to process exogenous antigens to direct them to the MHC I pathway (den Haan *et al.*, 2000; Pooley *et al.*, 2001) and/or on the ability of this DC subset to internalize cell-associated antigens (Schulz and Reis e Sousa, 2002).

Data derived from different experimental systems support the concept that CD8$^+$ DCs are efficient producers of interleukin (IL)-12, and are mainly involved in the induction of Th1 responses (reviewed in Iwasaki and Medzhitov, 2004). In this sense, it has been proposed that CD8$^+$ DCs play an essential role in the induction of CD8$^+$ T cell responses, particularly during infection by both cytolytic and noncytolytic viruses (Belz *et al.*, 2004, 2005), although recent data indicate that they are also involved in the initiation of T cell immunity against intracellular bacteria (Belz *et al.*, 2005). On the other hand, CD8$^+$ DC involvement in the induction of anti-protozoan parasite immunity is controversial. Whereas CD8$^+$ DCs have been demonstrated to play an essential role in the induction of Th1 responses against *Toxoplasma gondii* (Aliberti *et al.*, 2000), they do not appear to participate in anti-*Leishmania major* Th1 protective responses (Filippi *et al.*, 2003; Ritter *et al.*, 2004).

CD8$^-$ DCs

In contrast to CD8$^+$ DCs, CD8$^-$ DCs appear to be mainly located adjacent to the principal antigen entry zones of the lymphoid organs, i.e. the outer part of the splenic white pulp in association with the marginal zone, the subsinusal layer beneath the lymph node subcapsular sinus and the subepithelial area of the lymphoid tissue of the intestinal tract (reviewed by Villadangos and Heath, 2005). In the spleen, around two-thirds of CD8$^-$ DCs are CD4$^+$, whereas in the Peyer's patches and both in the peripheral and mesenteric lymph nodes the vast majority of CD8$^-$ DCs correspond to the CD4$^-$ subset (reviewed by Johansson and Kelsall, 2005). However, it has been reported that during infection by the mouse mammary tumor virus, there was a progressive increase in the proportion of CD4$^+$ DCs among the CD8$^-$ DC subpopulation located in the popliteal lymph nodes draining the infection area (Martin *et al.*, 2002).

CD8$^-$ DCs display a high endocytic and phagocytic capacity, but are inefficient in capturing apoptotic cells. In this sense, it has been reported that CD8$^-$ DCs are responsible for the in vivo stimulation of CD4$^+$ T cells after immunization with antigen-loaded DC-derived exosomes (Thery *et al.*, 2002). Data on the role of CD8$^-$ DCs in the induction of T cell responses in vivo are limited, and in particular those addressing differentially the function of CD8$^-$ CD4$^+$ DCs versus CD8$^-$ CD4$^-$ DCs. Globally, results derived from in vivo, but principally in vitro experiments performed with splenic CD8⁻ DCs support the concept that, in contrast to their CD8⁺ counterparts, CD8⁻ DCs appear to be poor IL-12 producers, and would be mainly involved in the induction of Th2 responses (reviewed in Iwasaki and Medzhitov, 2004). In addition, CD8⁻ DCs from the Peyer's patches (belonging to the CD4⁻ subset) have been demonstrated to produce high amounts of IL-10, and to induce Th2 responses leading to IgA production (Iwasaki and Kelsall, 1999; Sato et al., 2003). Moreover, these cells can also induce the differentiation of IL-10 producing T regulatory cells (Iwasaki and Kelsall, 2001). However the Th1 versus Th2-polarizing potential of CD8⁻ DCs remains controversial, and therefore needs to be addressed carefully, because it has been reported that CD8⁻ DCs can produce IL-12 in the presence of anti-IL-10 (Fallarino et al., 2002; Maldonado-Lopez et al., 2001) in response to α-galactosylcer-amide (Fujii et al., 2003) and Toll-like receptors (TLR)-7 ligands (Edwards et al., 2003).

On the other hand, a role for CD8⁻ DCs in the induction of protective immunity against *Leishmania major* has been proposed by Filippi *et al.* (2003), who reported that CD11b⁺ CD8⁻ DCs located in the lymph nodes draining the site of infection were responsible for the induction of Lack-specific Th1 responses. However, a more detailed phenotypic and functional analysis of these DCs is required to demonstrate conclusively whether these CD11b⁺ CD8⁻ DCs actually correspond to those CD8⁻ DCs pre-existing in the lymph node prior to the onset of *Leishmania* infection. Alternatively, these CD11b⁺ CD8⁻ DCs could represent dermal DCs that had migrated from the skin as proposed by Iwasaki (2003), or to newly-formed monocyte-derived DCs that can be also described as CD11b⁺ CD8⁻ DCs. In this sense, dermal DCs and monocyte-derived DCs, but not CD8⁻ DCs, express DEC-205 (Henri et al., 2001) and Ly-6C (C. Ardavín et al., unpublished data), respectively.

These data suggest that under certain physiological conditions CD8⁻ DCs could participate in the induction of Th1 responses. On the other hand, discrepancy between published data could rely on the yet undefined differential functions of the CD4⁺ versus CD4⁻ subsets of CD8⁻ DCs, because the majority of the experiments addressing the role of CD8⁻ DCs have been carried out with unfractionated CD8⁻ DCs.

CD11b⁻ DCs

CD11b⁻ DCs are preferentially found in the Peyer's patches and mesenteric lymph nodes, and represent only a minor subset in the spleen and peripheral lymph nodes; in addition they are the principal DC subpopulation in the liver (reviewed by Johansson and Kelsall, 2005). Within the Peyer's patches,

CD11b$^-$ DCs are located both in the subepithelial dome and the inter-follicular regions; in addition immature CD11b$^-$ DCs appear to be present within the follicular-associated epithelium. In the mesenteric lymph nodes CD11b$^-$ DCs are located in the T cell areas.

In contrast to IL-10 producing intestinal CD8$^-$ DCs, CD11b$^-$ produced IL-12 after microbial stimulation and induce Th1-polarized responses (Iwasaki and Kelsall, 2001). In addition, recent data from Dr. Kelsall's laboratory demonstrate that CD11b$^-$ DCs can internalize and process antigens from type 1 Reovirus-infected Peyer's patch epithelial cells, and stimulate virus-specific T cells by cross-priming (Fleeton *et al.*, 2004).

1.1.2 Migrating cDCs

Langerhans cells, dermal DCs and other interstitial DC subpopulations are strategically located in close association with epithelial surfaces that are exposed to microbial pathogens. They constitute a group of DC subpopula-tions expressing specialized endocytic and phagocytic receptors allowing a highly efficient uptake and processing of pathogen-derived antigens (reviewed by Valladeau and Saeland, 2005). In addition, it has been demonstrated that lamina propria-specific DCs can extend dendrites between epithelial cells and sample microorganisms from the intestinal lumen (Rescigno *et al.*, 2001).

Recent data suggest that migrating-cDC subpopulations display a constitutive rate of migration to the lymph nodes after antigen uptake in the absence of inflammatory signals that is generally associated with the induction of T cell tolerance. However, whereas this concept appears to be true for dermal or interstitial DCs (Itano *et al.*, 2003; Scheinecker *et al.*, 2002; Turley *et al.*, 2003), a recent study supports that Langerhans cells can activate naive T cells upon migration to the lymph nodes, in the steady state (Mayerova *et al.*, 2004). Nevertheless, as described for CD8$^+$ DCs and CD8$^-$ DCs, if antigen capture occurs in the presence of inflammatory stimuli such as TNF-α or IL-1β, or TLR-ligands, migrating-cDCs undergo a CCR7-dependent process of migration accompanied by maturation that leads to the induction of T cell immunogenic responses in the draining lymph nodes (reviewed by Valladeau and Saeland, 2005). However, although a series of reports support the view that both Langerhans cells and dermal or interstitial DCs can fulfill this function, recent data challenge this paradigm of DC immunobiology. In this sense, experimental evidence based on the induction of T cell responses in vivo suggest that Langerhans cells may not have an important role in the capture, transport and presentation of skin

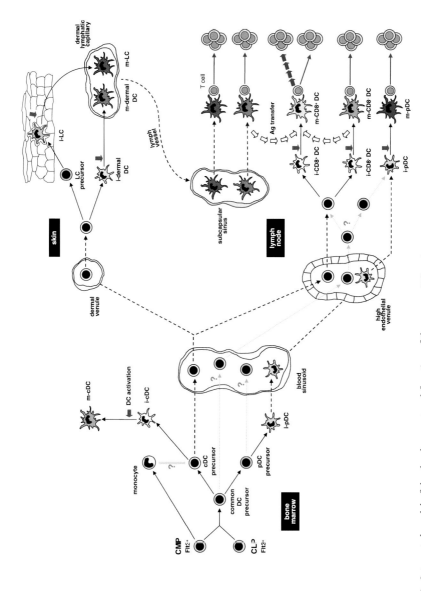

Figure 1.1. Integrated model of the development and function of the mouse DC system.

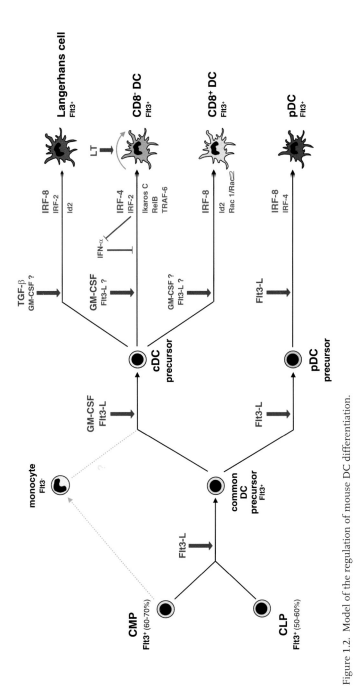

Figure 1.2. Model of the regulation of mouse DC differentiation.

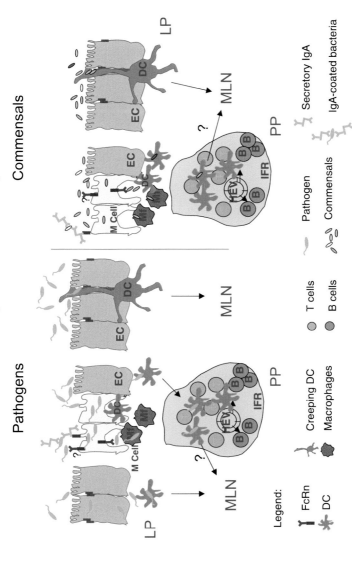

Figure 4.1. Mechanisms of bacterial uptake.

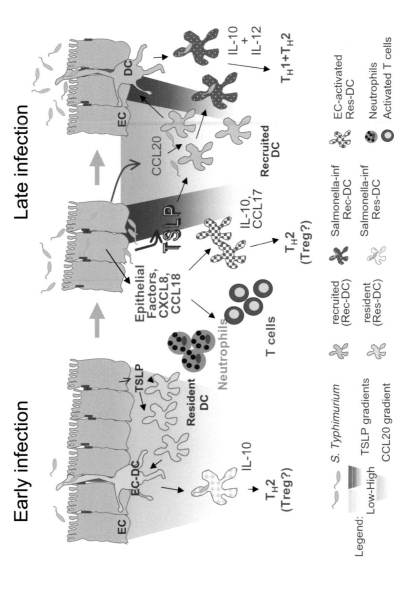

Figure 4.2. The cross-talk between ECs and DCs in *Salmonella typhimurium* handling.

pathogen-derived antigens to T cells, despite their capacity to internalize a large variety of microbial pathogens (reviewed by Villadangos and Heath, 2005). This function would be rather carried out essentially by dermal DCs or interstitial cDCs, as demonstrated during infection by *Leishmania major* (Ritter *et al.*, 2004) or *Herpes simplex* virus type 2 (Iwasaki, 2003). On the other hand, it has been reported that in certain viral infection models, neither dermal nor interstitial cDCs that capture and transport antigens to the lymph nodes, have an essential role in the induction of $CD8^+$ T cell responses. In this sense, it has been demonstrated that the $CD8^+$ T cell response against *Herpes simplex* virus type 1 was mediated by $CD8^+$ DCs, to which the antigen was transferred from interstitial-cDCs upon migration to the lymph node (Allan *et al.*, 2003). In a different experimental system, it has been shown that during infection by *Influenza* or *Herpes simplex* virus type 1, virus-derived antigens were presented to specific $CD8^+$ T cells in the draining lymph nodes, by both interstitial migrating-cDCs that have internalized the virus in the lung, and by resident lymph node $CD8^+$ DCs, to which migrating-cDCs had transferred the antigen (Belz *et al.*, 2004).

1.1.3 Plasmacytoid DCs

Murine pDCs can be characterized as $CD11c^+$ $CD11b^-$ $B220^+$ $Ly-6C^+$ cells, and are found in the bone marrow, thymus, blood and all peripheral lymphoid organs. In contrast to cDCs, pDCs appear to differentiate in the bone marrow (at least under non-inflammatory and/or infectious conditions), and express CD62L (L-selectin) that allows them to be recruited through high endothelial venules to the lymph nodes and the lymphoid tissue associated to the intestinal and respiratory tracts (reviewed by Barchet *et al.*, 2005). Interestingly, three pDC-specific monoclonal antibodies (mAbs) have been developed recently and represent powerful tools to analyze the development and function of pDCs. The mAb 120G8 (Asselin-Paturel *et al.*, 2003) recognizes a molecule that is expressed specifically by pDCs in non-stimulated mice, but is upregulated by type I interferon (IFN) on other cell types. The mAb 440c mAb (Blasius *et al.*, 2004) specifically recognizes pDCs in both naive and type I IFN-stimulated mice; interestingly pDCs produce type I IFN after 440c-mediated engagement of its yet undefined counter-receptor. Finally the mAb mPDCA-1 (Krug *et al.*, 2004) specifically recognizes pDCs. Treatment by either mPDCA-1 or 120G8 promotes pDC depletion in vivo. On the other hand, based on the expression of Ly49Q, two pDC subsets have been recently defined (Kamogawa-Schifter *et al.*, 2005). Peripheral pDCs are $Ly49Q^+$, whereas both $Ly49Q^+$ and $Ly49Q^-$ pDCs exist in bone marrow, the latter being less efficient in producing inflammatory cytokines, such as IL-6, IL-12 and TNF-α.

pDCs are characterized by their capacity to produce large amounts of type I-IFN during viral infections that relies essentially in a TLR-dependent mechanism initiated after virus endocytosis, and subsequent recognition of viral nucleic acids by TLR-7 or TLR-9 in an acidic endosomal compartment (Iwasaki and Medzhitov, 2004). The high type I IFN production efficiency of pDCs appears to be in part related to their constitutive expression of interferon regulatory factor (IRF)-7, allowing its rapid activation and subsequent transcription of IFN-α genes, independently of previous IRF-3 activation (as described for IRF-7 induction in non-pDCs) (Barchet *et al.*, 2002; Prakash *et al.*, 2005). Differential transport of viral nucleic acids into endosomal versus lysosomal compartments in pDCs and cDCs, respectively, could also contribute to the functional specialization of pDCs as type I IFN producers.

Apart from providing this type I IFN-mediated initial defense mechanism against viral infections after exposure to DC maturation stimuli, such as TLR-7/9 ligands or CD40L, pDCs could also participate in the induction of antigen-specific T cell responses (reviewed by Colonna *et al.*, 2004), although whether pDCs can activate naïve T cells in vivo after viral infection remains controversial. However, pDCs have been demonstrated to induce the proliferation of virus-specific, antigen-primed CD8$^+$ T cells and Th1 CD4$^+$ T cells (Krug *et al.*, 2003; Schlecht *et al.*, 2004). Based on these data, it has been hypothesized that pDCs could synergize with cDCs in the induction of Th1 responses during in vivo viral infections.

1.2 DENDRITIC CELL DIFFERENTIATION

Current knowledge on the origin and differentiation pathways leading to the generation of DCs relies essentially on in vivo DC reconstitution experiments in which defined hematopoietic precursors were transferred into irradiated mice, and on in vivo and in vitro studies performed with mice deficient in growing or transcription factors involved in DC differentiation (reviewed by Ardavín, 2003). Due in part to the technical complexity and limitations of such assays, the majority of these studies have focused almost exclusively on the differentiation of CD8$^+$ DCs, CD8$^-$ DCs and pDCs, and occasionally on skin Langerhans cells. As a consequence, how the differentiation of other DC subpopulations proceeds remains essentially unknown. Nevertheless, as discussed below, these experiments have provided valuable information on the differentiation pathways generating DCs from early myeloid or lymphoid precursors. However, unfortunately they have so far not allowed a comprehensive characterization of the immediate precursors of DCs that home to specific effector sites where DCs exert their function. Figure 1.1 summarizes current knowledge on the development of DC subpopulations.

1.2.1 Myeloid and lymphoid DC differentiation pathways

DCs were originally thought to be derived exclusively from myeloid precursors, but experiments showing that both thymic and splenic DCs (particularly the $CD8^+$ and $CD8^-$ DC subpopulations) can be originated from early thymic precursors, devoid of myeloid reconstitution potential (Ardavín et al., 1993), led to the hypothesis that DCs could be also generated through a lymphoid differentiation pathway. This theory was reinforced by additional data from experiments demonstrating that bone marrow common lymphoid progenitors (CLPs) had the capacity to differentiate into the same DC subpopulations as those generated experimentally by early thymic precursors (Traver et al., 2000). Subsequent studies, performed by co-transferring CLPs and common myeloid progenitors (CMPs) into irradiated recipients, indicated that, at least under these experimental conditions, the differentiation of $CD8^+$ DCs and $CD8^-$ DCs in the thymus and spleen was the result of a simultaneous contribution of CLPs and CMPs (Manz et al., 2001; Wu et al., 2001). The concept that certain DC subpopulations can be generated through both the myeloid and the lymphoid differentiation pathways was recently extended to the pDC subpopulation, by demonstrating that both CMPs and CLPs can give rise to pDCs after intravenous transfer (D'Amico and Wu, 2003) or in Flt3L-driven cultures (Karsunky et al., 2005).

Although so far the developmental derivation of organ-specific DC subpopulations has not been addressed in depth, globally these data support the hypothesis that both myeloid and lymphoid precursors are endowed with the capacity to differentiate into diverse DC subpopulations. However, whether these experiments reflect the physiological situation, i.e. whether in fact both differentiation pathways actually contribute to the generation of the different DC subpopulations, and if so, what is the differential contribution of CMPs and CLPs to the generation of each DC subpopulation, remains essentially unknown. In this regard, the model of the dual contribution of the myeloid and lymphoid pathways to DC development appears to be particularly controversial with regard to the differentiation of pDCs. This DC subpopulation has been claimed to derive from lymphoid precursors under physiological conditions, based on a series of studies performed both in humans and mice, describing that pDCs express several molecules closely related to the lymphoid lineage, such as pTα, spi-B, IL-7R, PIII CIITA, and have IgH gene rearrangements (reviewed by Colonna *et al.*, 2004). However, these data do not allow one to draw a definitive conclusion on pDC origin because two recently published reports support the view that in fact human and mouse pDCs can activate a lymphoid genetic program irrespective of whether they are derived from lymphoid or myeloid progenitors (Chicha *et al.*, 2004; Shigematsu *et al.*, 2004).

1.2.2 Control of DC differentiation

Interestingly, the fact that the DC differentiation capacity of CMPs and CLPs is largely associated with the Flt3-positive fraction of these progenitor populations (D'Amico and Wu, 2003) suggests that the DC differentiation potential of myeloid and lymphoid hematopoietic precursors is determined by the expression of the receptor for Flt3L. The cytokine Flt3L plays a crucial role in DC development, as demonstrated in Flt3L-deficient mice displaying a profound defect in the differentiation of all the DC subpopulations that were analyzed (McKenna *et al.*, 2000). Although both in vivo and in vitro evidence indicate that Flt3L determines the commitment of CMPs and CLPs toward DC differentiation, and drives the early steps of this process (reviewed by Ardavín, 2003), a number of experimental evidences suggest that additional cytokines are selectively required to induce and direct the generation of specific DC subpopulations. Unfortunately, data currently available on this issue are far from allowing the establishment of a comprehensive model of cytokine-mediated regulation of the differentiation of mouse DC subpopulations. In this regard, although as discussed below, a number of cytokines appear to be involved in defined pathways of DC differentiation, the cytokine combinations controlling the development of each DC subpopulation remain to be defined.

In this regard, Flt3L is not only a key cytokine in driving the initial steps of DC differentiation from CMPs and CLPs, but also appears to directly control the generation of pDCs (Karsunky *et al.*, 2005; Laouar *et al.*, 2003). Granulocyte-macrophage colony-stimulating factor (GM-CSF) has been demonstrated to drive the differentiation of mouse monocytes into DCs (Leon *et al.*, 2005), but its precise role on the generation of the different DC subpopulations has not been addressed in depth. However, on the basis of data derived from experiments of in vivo treatment with GM-CSF (Daro *et al.*, 2000), and from the analysis of IRF-4 and/or IRF-8-deficient mice (Tamura *et al.*, 2005), it has been proposed that GM-CSF and Flt3-L control, through IRF-4 and IRF-8, the differentiation of CD8$^-$ DCs and CD8$^+$

DCs, respectively. Moreover, GM-CSF appears to inhibit the differentiation of pDCs (Gilliet *et al.*, 2002). Mice deficient in LTα, or in its receptor LTβ-R, display a deficient splenic CD8⁻ DC development, which has been correlated with the ability of LTα to induce the proliferation of CD8⁻ DCs (Kabashima *et al.*, 2005). Finally, both in vivo (Borkowski *et al.*, 1996) and in vitro (Zhang *et al.*, 1999) assays have demonstrated the requirement of TGF-β for the differentiation of epidermal Langerhans cells.

Studies published over the last few years have provided new insights on the signaling pathways and transcriptional regulation associated with the development of certain DC subpopulations (reviewed by Ardavín, 2003). Several members of the IRF-family of transcription factors have been proposed to control the differentiation of diverse DC subpopulations. In this regard, the analysis of mice deficient in IRF-8 has demonstrated the involvement of this transcription factor in the differentiation of CD8⁺ DCs (Aliberti *et al.*, 2003; Schiavoni *et al.*, 2002), Langerhans cells (Schiavoni *et al.*, 2004) and pDCs (Schiavoni *et al.*, 2002; Tsujimura *et al.*, 2003), whereas IRF-4 is required for the development of CD8⁻ CD4⁺ splenic DCs (Suzuki *et al.*, 2004), and to a lesser extent of pDCs (Tamura *et al.*, 2005). Finally, IRF-2 has been shown to participate in the differentiation of Langerhans cells (Ichikawa *et al.*, 2004), and splenic CD8⁻ CD4⁺ DCs (Honda *et al.*, 2004; Ichikawa *et al.*, 2004), whose dependence on IRF-2 has been claimed to rely on the inhibition of IFN-α production that would in turn negatively control the generation of CD8⁻ CD4⁺ DCs.

In addition, a defective development of CD8⁻ DCs has also been reported in mice deficient in the adapter molecule TRAF-6 (involved in the signaling pathway of IL-1, TNF-α and TLRs) (Kobayashi *et al.*, 2003), and in the transcription factors Ikaros C (Wu *et al.*, 1997) and RelB (Wu *et al.*, 1998). Finally, the differentiation of splenic CD8⁺ DCs appears to require the Rho family guanosine triphosphatases Rac 1 and Rac 2 (involved in DC migration) (Benvenuti *et al.*, 2004), and the transcription factor Id2 (Hacker *et al.*, 2003), the latter being also involved in the differentiation of epidermal Langerhans cells. Based on these data, a model of the control of mouse DC differentiation is proposed in Figure 1.2.

Despite the relevance of these results, the specific cytokines that initiate these signaling pathways and activate these transcription factors on immediate DC precursors have to be precisely defined. Based on the current literature on this subject, it can be hypothesized that the generation of the different DC subpopulations is controlled not only by cytokines that trigger specific differentiation programs on immediate DC precursors, but also by cytokines and chemokines responsible for the correct

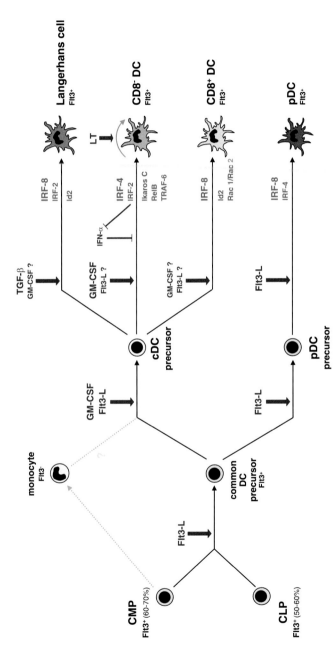

Figure 1.2. Model of the regulation of mouse DC differentiation. This figure summarizes the current knowledge on the regulation of mouse DC differentiation. This figure summarizes the current knowledge on the principal cytokines (in blue), and transcription factors and adapter molecules (in red) claimed to control the different DC differentiation pathways. (For a colour version of this figure, please refer to the colour insert between pages 12 and 13.)

development of defined DC subpopulations, by allowing immediate DC precursors to locate in specific environments. In this regard, future research on DC development and acquisition of DC functional specialization needs to be focused on the characterization of DC precursors because, although recent research on DC development has provided important insights on the origin and control of the differentiation pathways leading to the different DC subpopulations, the identity of the immediate DC precursors and how their differentiation is controlled has only begun to be unravelled.

REFERENCES

Aliberti, J., Reis e Sousa, C., Schito, M., Hieny, S., Wells, T., Huffnagle, G. B. and Sher, A. (2000). CCR5 provides a signal for microbial induced production of IL-12 by CD8 alpha+ dendritic cells. *Nat. Immunol.*, **1**, 83–7.

Aliberti, J., Schulz, O., Pennington, D. J., Tsujimura, H., Reis e Sousa, C., Ozato, K. and Sher, A. (2003). Essential role for ICSBP in the in vivo development of murine CD8alpha+ dendritic cells. *Blood*, **101**, 305–10.

Allan, R. S., Smith, C. M., Belz, G. T., van Lint, A. L., Wakim, L. M., Heath, W. R. and Carbone, F. R. (2003). Epidermal viral immunity induced by CD8alpha+ dendritic cells but not by Langerhans cells. *Science*, **301**, 1925–8.

Ardavín, C. (1997). Thymic dendritic cells. *Immunol Today*, **18**, 350–61.

Ardavín, C. (2003). Origin, precursors and differentiation of mouse dendritic cells. *Nat. Rev. Immunol.*, **3**, 582–90.

Ardavín, C., Wu, L., Li, C. L. and Shortman, K. (1993). Thymic dendritic cells and T cells develop simultaneously in the thymus from a common precursor population. *Nature*, **362**, 761–3.

Asselin-Paturel, C., Brizard, G., Pin, J. J., Briere, F. and Trinchieri, G. (2003). Mouse strain differences in plasmacytoid dendritic cell frequency and function revealed by a novel monoclonal antibody. *J. Immunol.*, **171**, 6466–77.

Barchet, W., Cella, M., Odermatt, B., Asselin-Paturel, C., Colonna, M. and Kalinke, U. (2002). Virus-induced interferon alpha production by a dendritic cell subset in the absence of feedback signaling in vivo. *J. Exp. Med.*, **195**, 507–16.

Barchet, W., Cella, M. and Colonna, M. (2005). Plasmacytoid dendritic cells-virus experts of innate immunity. *Semin. Immunol.*, **17**, 253–61.

Belz, G. T., Behrens, G. M., Smith, C. M., Miller, J. F., Jones, C., Lejon, K., Fathman, C. G., Mueller, S. N., Shortman, K., Carbone, F. R. and Heath, W. R. (2002). The CD8alpha(+) dendritic cell is responsible for inducing peripheral self-tolerance to tissue-associated antigens. *J. Exp. Med.*, **196**, 1099–104.

Belz, G. T., Smith, C. M., Kleinert, L., Reading, P., Brooks, A., Shortman, K., Carbone, F. R. and Heath, W. R. (2004). Distinct migrating and nonmigrating dendritic cell populations are involved in MHC class I-restricted antigen presentation after lung infection with virus. *Proc. Natl Acad. Sci. U S A*, **101**, 8670–5.

Belz, G. T., Shortman, K., Bevan, M. J. and Heath, W. R. (2005). CD8alpha+ dendritic cells selectively present MHC class I-restricted noncytolytic viral and intracellular bacterial antigens in vivo. *J. Immunol.*, **175**, 196–200.

Benvenuti, F., Hugues, S., Walmsley, M., Ruf, S., Fetler, L., Popoff, M., Tybulewicz, V. L. and Amigorena, S. (2004). Requirement of Rac1 and Rac2 expression by mature dendritic cells for T cell priming. *Science*, **305**, 1150–3.

Blasius, A., Vermi, W., Krug, A., Facchetti, F., Cella, M. and Colonna, M. (2004). A cell-surface molecule selectively expressed on murine natural interferon-producing cells that blocks secretion of interferon-alpha. *Blood*, **103**, 4201–6.

Borkowski, T. A., Letterio, J. J., Farr, A. G. and Udey, M. C. (1996). A role for endogenous transforming growth factor beta 1 in Langerhans cell biology: the skin of transforming growth factor beta 1 null mice is devoid of epidermal Langerhans cells. *J. Exp. Med.*, **184**, 2417–22.

Chicha, L., Jarrossay, D. and Manz, M. G. (2004). Clonal type I interferon-producing and dendritic cell precursors are contained in both human lymphoid and myeloid progenitor populations. *J. Exp. Med.*, **200**, 1519–24.

Colonna, M., Trinchieri, G. and Liu, Y. J. (2004). Plasmacytoid dendritic cells in immunity. *Nat. Immunol.*, **5**, 1219–26.

D'Amico, A. and Wu, L. (2003). The early progenitors of mouse dendritic cells and plasmacytoid predendritic cells are within the bone marrow hemopoietic precursors expressing Flt3. *J. Exp. Med.*, **198**, 293–303.

Daro, E., Pulendran, B., Brasel, K., Teepe, M., Pettit, D., Lynch, D. H., Vremec, D., Robb, L., Shortman, K., McKenna, H. J., Maliszewski, C. R. and Maraskovsky, E. (2000). Polyethylene glycol-modified GM-CSF expands CD11b(high)CD11c(high) but not CD11b(low)CD11c(high) murine dendritic cells in vivo: a comparative analysis with Flt3 ligand. *J. Immunol.*, **165**, 49–58.

De Smedt, T., Pajak, B., Muraille, E., Lespagnard, L., Heinen, E., De Baetselier, P., Urbain, J., Leo, O. and Moser, M. (1996). Regulation of dendritic cell numbers and maturation by lipopolysaccharide in vivo. *J. Exp. Med.*, **184**, 1413–24.

den Haan, J. M., Lehar, S. M. and Bevan, M. J. (2000). CD8(+) but not CD8(−) dendritic cells cross-prime cytotoxic T cells in vivo. *J. Exp. Med.*, **192**, 1685–96.

Edwards, A. D., Diebold, S. S., Slack, E. M., Tomizawa, H., Hemmi, H., Kaisho, T., Akira, S. and Reis e Sousa, C. (2003). Toll-like receptor expression in murine DC subsets: lack of TLR7 expression by CD8 alpha+ DC correlates with unresponsiveness to imidazoquinolines. *Eur. J. Immunol.*, **33**, 827–33.

Fallarino, F., Grohmann, U., Vacca, C., Bianchi, R., Fioretti, M. C. and Puccetti, P. (2002). CD40 ligand and CTLA-4 are reciprocally regulated in the Th1 cell proliferation response sustained by CD8(+) dendritic cells. *J. Immunol.*, **169**, 1182–8.

Filippi, C., Hugues, S., Cazareth, J., Julia, V., Glaichenhaus, N. and Ugolini, S. (2003). CD4+ T cell polarization in mice is modulated by strain-specific major histocompatibility complex-independent differences within dendritic cells. *J. Exp. Med.*, **198**, 201–9.

Fleeton, M. N., Contractor, N., Leon, F., Wetzel, J. D., Dermody, T. S. and Kelsall, B. L. (2004). Peyer's patch dendritic cells process viral antigen from apoptotic epithelial cells in the intestine of reovirus-infected mice. *J. Exp. Med.*, **200**, 235–45.

Fujii, S., Shimizu, K., Smith, C., Bonifaz, L. and Steinman, R. M. (2003). Activation of natural killer T cells by alpha-galactosylceramide rapidly induces the full maturation of dendritic cells in vivo and thereby acts as an adjuvant for combined CD4 and CD8 T cell immunity to a coadministered protein. *J. Exp. Med.*, **198**, 267–79.

Gilliet, M., Boonstra, A., Paturel, C., Antonenko, S., Xu, X. L., Trinchieri, G., O'Garra, A. and Liu, Y. J. (2002). The development of murine plasmacytoid dendritic cell precursors is differentially regulated by FLT3-ligand and granulocyte/macrophage colony-stimulating factor. *J. Exp. Med.*, **195**, 953–8.

Hacker, C., Kirsch, R. D., Ju, X. S., Hieronymus, T., Gust, T. C., Kuhl, C., Jorgas, T., Kurz, S. M., Rose-John, S., Yokota, Y. and Zenke, M. (2003). Transcriptional profiling identifies Id2 function in dendritic cell development. *Nat. Immunol.*, **4**, 380–6.

Henri, S., Vremec, D., Kamath, A., Waithman, J., Williams, S., Benoist, C., Burnham, K., Saeland, S., Handman, E. and Shortman, K. (2001). The dendritic cell populations of mouse lymph nodes. *J. Immunol.*, **167**, 741–8.

Honda, K., Mizutani, T. and Taniguchi, T. (2004). Negative regulation of IFN-alpha/beta signaling by IFN regulatory factor 2 for homeostatic development of dendritic cells. *Proc. Natl Acad. Sci. U S A*, **101**, 2416–21.

Ichikawa, E., Hida, S., Omatsu, Y., Shimoyama, S., Takahara, K., Miyagawa, S., Inaba, K. and Taki, S. (2004). Defective development of splenic and epidermal CD4+ dendritic cells in mice deficient for IFN regulatory factor-2. *Proc. Natl Acad. Sci. U S A*, **101**, 3909–14.

Itano, A. A., McSorley, S. J., Reinhardt, R. L., Ehst, B. D., Ingulli, E., Rudensky, A. Y. and Jenkins, M. K. (2003). Distinct dendritic cell populations sequentially present antigen to CD4 T cells and stimulate different aspects of cell-mediated immunity. *Immunity*, **19**, 47–57.

Iwasaki, A. (2003). The importance of CD11b+ dendritic cells in CD4+ T cell activation in vivo: with help from interleukin 1. *J. Exp. Med.*, **198**, 185–90.

Iwasaki, A. and Kelsall, B. L. (1999). Freshly isolated Peyer's patch, but not spleen, dendritic cells produce interleukin 10 and induce the differentiation of T helper type 2 cells. *J. Exp. Med.*, **190**, 229–39.

Iwasaki, A. and Kelsall, B. L. (2000). Localization of distinct Peyer's patch dendritic cell subsets and their recruitment by chemokines macrophage inflammatory protein (MIP)-3alpha, MIP-3beta, and secondary lymphoid organ chemokine. *J. Exp. Med.*, **191**, 1381–94.

Iwasaki, A. and Kelsall, B. L. (2001). Unique functions of CD11b+, CD8 alpha+, and double-negative Peyer's patch dendritic cells. *J. Immunol.*, **166**, 4884–90.

Iwasaki, A. and Medzhitov, R. (2004). Toll-like receptor control of the adaptive immune responses. *Nat. Immunol.*, **5**, 987–95.

Iyoda, T., Shimoyama, S., Liu, K., Omatsu, Y., Akiyama, Y., Maeda, Y., Takahara, K., Steinman, R. M. and Inaba, K. (2002). The CD8+ dendritic cell subset selectively endocytoses dying cells in culture and in vivo. *J. Exp. Med.*, **195**, 1289–302.

Johansson, C. and Kelsall, B. L. (2005). Phenotype and function of intestinal dendritic cells. *Semin. Immunol.*, **17**, 284–94.

Kabashima, K., Banks, T. A., Ansel, K. M., Lu, T. T., Ware, C. F. and Cyster, J. G. (2005). Intrinsic lymphotoxin-beta receptor requirement for homeostasis of lymphoid tissue dendritic cells. *Immunity*, **22**, 439–50.

Kamogawa-Schifter, Y., Ohkawa, J., Namiki, S., Arai, N., Arai, K. and Liu, Y. (2005). Ly49Q defines 2 pDC subsets in mice. *Blood*, **105**, 2787–92.

Karsunky, H., Merad, M., Mende, I., Manz, M. G., Engleman, E. G. and Weissman, I. L. (2005). Developmental origin of interferon-alpha-producing dendritic cells from hematopoietic precursors. *Exp. Hematol.*, **33**, 173–81.

Kobayashi, T., Walsh, P. T., Walsh, M. C., Speirs, K. M., Chiffoleau, E., King, C. G., Hancock, W. W., Caamano, J. H., Hunter, C. A., Scott, P., Turka, L. A. and Choi, Y. (2003). TRAF6 is a critical factor for dendritic cell maturation and development. *Immunity*, **19**, 353–63.

Krug, A., Veeraswamy, R., Pekosz, A., Kanagawa, O., Unanue, E. R., Colonna, M. and Cella, M. (2003). Interferon-producing cells fail to induce proliferation of naive T cells but can promote expansion and T helper 1 differentiation of antigen-experienced unpolarized T cells. *J. Exp. Med.*, **197**, 899–906.

Krug, A., French, A. R., Barchet, W., Fischer, J. A., Dzionek, A., Pingel, J. T., Orihuela, M. M., Akira, S., Yokoyama, W. M. and Colonna, M. (2004). TLR9-dependent recognition of MCMV by IPC and DC generates coordinated cytokine responses that activate antiviral NK cell function. *Immunity*, **21**, 107–19.

Laouar, Y., Welte, T., Fu, X. Y. and Flavell, R. A. (2003). STAT3 is required for Flt3L-dependent dendritic cell differentiation. *Immunity*, **19**, 903–12.

Leenen, P. J., Radosevic, K., Voerman, J. S., Salomon, B., van Rooijen, N., Klatzmann, D. and van Ewijk, W. (1998). Heterogeneity of mouse spleen dendritic cells: in vivo phagocytic activity, expression of macrophage markers, and subpopulation turnover. *J. Immunol.*, **160**, 2166–73.

Leon, B., Lopez-Bravo, M. and Ardavín, C. (2005). Monocyte-derived dendritic cells. *Semin. Immunol.*, **17**, 313–18.

Liu, K., Iyoda, T., Saternus, M., Kimura, Y., Inaba, K. and Steinman, R. M. (2002). Immune tolerance after delivery of dying cells to dendritic cells in situ. *J. Exp. Med.*, **196**, 1091–7.

Maldonado-Lopez, R., Maliszewski, C., Urbain, J. and Moser, M. (2001). Cytokines regulate the capacity of CD8alpha(+) and CD8alpha(−) dendritic cells to prime Th1/Th2 cells in vivo. *J. Immunol.*, **167**, 4345–50.

Manz, M. G., Traver, D., Miyamoto, T., Weissman, I. L. and Akashi, K. (2001). Dendritic cell potentials of early lymphoid and myeloid progenitors. *Blood*, **97**, 3333–41.

Martin, P., Ruiz, S. R., del Hoyo, G. M., Anjuere, F., Vargas, H. H., Lopez-Bravo, M. and Ardavín, C. (2002). Dramatic increase in lymph node dendritic cell number during infection by the mouse mammary tumor virus occurs by a CD62L-dependent blood-borne DC recruitment. *Blood*, **99**, 1282–8.

Mayerova, D., Parke, E. A., Bursch, L. S., Odumade, O. A. and Hogquist, K. A. (2004). Langerhans cells activate naive self-antigen-specific CD8 T cells in the steady state. *Immunity*, **21**, 391–400.

McKenna, H. J., Stocking, K. L., Miller, R. E., Brasel, K., De Smedt, T., Maraskovsky, E., Maliszewski, C. R., Lynch, D. H., Smith, J., Pulendran, B., Roux, E. R., Teepe, M., Lyman, S. D. and Peschon, J. J. (2000). Mice lacking flt3 ligand have deficient hematopoiesis affecting hematopoietic progenitor cells, dendritic cells, and natural killer cells. *Blood*, **95**, 3489–97.

Pooley, J. L., Heath, W. R. and Shortman, K. (2001). Cutting edge: intravenous soluble antigen is presented to CD4 T cells by CD8-dendritic cells, but cross-presented to CD8 T cells by CD8+ dendritic cells. *J. Immunol.*, **166**, 5327–30.

Prakash, A., Smith, E., Lee, C. K. and Levy, D. E. (2005). Tissue-specific positive feedback requirements for production of type I interferon following virus infection. *J. Biol. Chem.*, **280**, 18651–7.

Reis e Sousa, C., Hieny, S., Scharton-Kersten, T., Jankovic, D., Charest, H., Germain, R. N. and Sher, A. (1997). In vivo microbial stimulation induces rapid CD40 ligand-independent production of interleukin 12 by dendritic cells and their redistribution to T cell areas. *J. Exp. Med.*, **186**, 1819–29.

Rescigno, M., Urbano, M., Valzasina, B., Francolini, M., Rotta, G., Bonasio, R., Granucci, F., Kraehenbuhl, J. P. and Ricciardi-Castagnoli, P. (2001). Dendritic cells express tight junction proteins and penetrate gut epithelial monolayers to sample bacteria. *Nat. Immunol.*, **2**, 361–7.

Ritter, U., Meissner, A., Scheidig, C. and Korner, H. (2004). CD8 alpha- and Langerin-negative dendritic cells, but not Langerhans cells, act as principal antigen-presenting cells in leishmaniasis. *Eur. J. Immunol.*, **34**, 1542–50.

Sato, A., Hashiguchi, M., Toda, E., Iwasaki, A., Hachimura, S. and Kaminogawa, S. (2003). CD11b+ Peyer's patch dendritic cells secrete IL-6 and induce IgA secretion from naive B cells. *J. Immunol.*, **171**, 3684–90.

Scheinecker, C., McHugh, R., Shevach, E. M. and Germain, R. N. (2002). Constitutive presentation of a natural tissue autoantigen exclusively by dendritic cells in the draining lymph node. *J. Exp. Med.*, **196**, 1079–90.

Schiavoni, G., Mattei, F., Sestili, P., Borghi, P., Venditti, M., Morse, H. C., 3rd, Belardelli, F. and Gabriele, L. (2002). ICSBP is essential for the development of mouse type I interferon-producing cells and for the generation and activation of CD8alpha(+) dendritic cells. *J. Exp. Med.*, **196**, 1415–25.

Schiavoni, G., Mattei, F., Borghi, P., Sestili, P., Venditti, M., Morse, H. C., 3rd, Belardelli, F. and Gabriele, L. (2004). ICSBP is critically involved in the normal development and trafficking of Langerhans cells and dermal dendritic cells. *Blood*, **103**, 2221–8.

Schlecht, G., Garcia, S., Escriou, N., Freitas, A. A., Leclerc, C. and Dadaglio, G. (2004). Murine plasmacytoid dendritic cells induce effector/memory CD8+ T-cell responses in vivo after viral stimulation. *Blood*, **104**, 1808–15.

Schulz, O. and Reis e Sousa, C. (2002). Cross-presentation of cell-associated antigens by CD8alpha+ dendritic cells is attributable to their ability to internalize dead cells. *Immunology*, **107**, 183–9.

Shigematsu, H., Reizis, B., Iwasaki, H., Mizuno, S., Hu, D., Traver, D., Leder, P., Sakaguchi, N. and Akashi, K. (2004). Plasmacytoid dendritic cells activate lymphoid-specific genetic programs irrespective of their cellular origin. *Immunity*, **21**, 43–53.

Shortman, K. and Liu, Y. J. (2002). Mouse and human dendritic cell subtypes. *Nat. Rev. Immunol.*, **2**, 151–61.

Steinman, R. M. and Nussenzweig, M. C. (2002). Avoiding horror autotoxicus: the importance of dendritic cells in peripheral T cell tolerance. *Proc. Natl Acad. Sci. U S A*, **99**, 351–8.

Suzuki, S., Honma, K., Matsuyama, T., Suzuki, K., Toriyama, K., Akitoyo, I., Yamamoto, K., Suematsu, T., Nakamura, M., Yui, K. and Kumatori, A. (2004). Critical roles of interferon regulatory factor 4 in CD11bhigh CD8alpha− dendritic cell development. *Proc. Natl Acad. Sci. U S A*, **101**, 8981–6.

Tamura, T., Tailor, P., Yamaoka, K., Kong, H. J., Tsujimura, H., O'Shea, J. J., Singh, H. and Ozato, K. (2005). IFN regulatory factor-4 and -8 govern dendritic cell subset development and their functional diversity. *J. Immunol.*, **174**, 2573–81.

Thery, C., Duban, L., Segura, E., Veron, P., Lantz, O. and Amigorena, S. (2002). Indirect activation of naive CD4$^+$ T cells by dendritic cell-derived exosomes. *Nat. Immunol.*, **3**, 1156–62.

Traver, D., Akashi, K., Manz, M., Merad, M., Miyamoto, T., Engleman, E. G. and Weissman, I. L. (2000). Development of CD8alpha-positive dendritic cells from a common myeloid progenitor. *Science*, **290**, 2152–4.

Tsujimura, H., Tamura, T. and Ozato, K. (2003). Cutting edge: IFN consensus sequence binding protein/IFN regulatory factor 8 drives the development of type I IFN-producing plasmacytoid dendritic cells. *J. Immunol.*, **170**, 1131–5.

Turley, S., Poirot, L., Hattori, M., Benoist, C. and Mathis, D. (2003). Physiological beta cell death triggers priming of self-reactive T cells by dendritic cells in a type-1 diabetes model. *J. Exp. Med.*, **198**, 1527–37.

Valladeau, J. and Saeland, S. (2005). Cutaneous dendritic cells. *Semin. Immunol.*, **17**, 273–83.

Villadangos, J. A. and Heath, W. R. (2005). Life cycle, migration and antigen presenting functions of spleen and lymph node dendritic cells: Limitations of the Langerhans cells paradigm. *Semin. Immunol.*, **17**, 262–72.

Wu, L., Nichogiannopoulou, A., Shortman, K. and Georgopoulos, K. (1997). Cell-autonomous defects in dendritic cell populations of Ikaros mutant mice point to a developmental relationship with the lymphoid lineage. *Immunity*, **7**, 483–92.

Wu, L., D'Amico, A., Winkel, K. D., Suter, M., Lo, D. and Shortman, K. (1998). RelB is essential for the development of myeloid-related CD8alpha− dendritic cells but not of lymphoid-related CD8alpha+ dendritic cells. *Immunity*, **9**, 839–47.

Wu, L., D'Amico, A., Hochrein, H., O'Keeffe, M., Shortman, K. and Lucas, K. (2001). Development of thymic and splenic dendritic cell populations from different hemopoietic precursors. *Blood*, **98**, 3376–82.

Zhang, Y., Zhang, Y. Y., Ogata, M., Chen, P., Harada, A., Hashimoto, S. and Matsushima, K. (1999). Transforming growth factor-beta1 polarizes murine hematopoietic progenitor cells to generate Langerhans cell-like dendritic cells through a monocyte/macrophage differentiation pathway. *Blood*, **93**, 1208–20.

CHAPTER 2

Toll-like receptor signaling

Osamu Takeuchi and Shizuo Akira
Osaka University

2.1 INTRODUCTION

Toll-like receptors (TLRs) play essential roles in innate immune responses[1,2]. The name TLR is derived from a *Drosophila* protein, Toll, which detects fungal infection in the fruit fly. The immune system in *Drosophila* is entirely dependent on a limited number of germline-encoded receptors for pathogen recognition. In contrast, the vertebrate immune system is characterized by the evolution of acquired immunity in addition to innate immunity. Acquired immunity is mediated by T and B cells, which utilize rearranged receptors. This system is advantageous for detecting pathogens with high specificity, eradicating infection in the late stages and establishing an immunological memory. However, the mammalian innate immune system plays critical roles in the initial defense against invading pathogens and subsequent activation of the acquired immune system. Innate immune cells, such as macrophages and dendritic cells (DCs), sense pathogens through TLRs, phagocytose them and evoke immune responses.

To date, 12 different TLRs have been reported in either humans or mice. The innate immunity system targets a set of molecular structures that are unique to microorganisms and shared by various pathogens, but absent from host cells. By recognizing these "pathogen-specific" patterns, the innate immunity system is able to prevent autoimmune responses. Members of the TLR family of proteins are characterized by extracellular leucine-rich repeat (LRR) motifs responsible for ligand recognition, a transmembrane region and a cytoplasmic tail containing a Toll/IL-1 receptor homology (TIR) domain. Extensive studies have identified the pathogenic components recognized by each TLR. TLR1, 2, 4, 5, 6 and 11 are involved in the recognition of microbial components that contain lipids, sugars and proteins. In detail, lipopolysaccharide (LPS) is recognized by TLR4[3,4], bacterial di-acyl lipoprotein is recognized by a heterodimer of TLR2 and

TLR6[5,6], tri-acyl lipoprotein is recognized by a heterodimer of TLR1 and TLR2[7,8], bacterial flagellin is recognized by TLR5[9] and uropathogenic bacteria and a profilin-like molecule from *Toxoplasma gondii* are detected by TLR11[10,11]. In contrast, TLR3, 7, 8 and 9 recognize nucleotides from pathogens. For example, double-stranded RNA (dsRNA) is recognized by TLR3[12], viral GU-rich single-stranded RNA (ssRNA) is recognized by TLR7 in mice and TLR8 in humans[13,14] and bacterial or viral unmethylated DNA with CpG motifs is recognized by TLR9[15–17]. TLR1, 2, 4, 5, 6 and 11 are expressed on the plasma membrane, whereas the TLRs that recognize nucleotides are localized to intracellular endosomes or lysosomes[18].

2.2 TLR SIGNALING AND TIR DOMAIN-CONTAINING ADAPTOR MOLECULES

Upon recognition of pathogenic components and ligand stimulation, TLRs activate similar, but distinct, signaling pathways via their TIR domains. These intracellular signaling pathways are responsible for various cellular and systemic immune responses, and have pleiotrophic effects including the production of proinflammatory cytokines, chemokines and interferons (IFNs), upregulation of surface co-stimulation molecule expression, phago-cytosis of bacteria and induction of B cell proliferation. DCs play pivotal roles in T cell activation and differentiation.

The initial step of TLR signaling is the recruitment of cytoplasmic adaptor molecules to the receptor. These adaptors possess a TIR domain and associate with the TIR domain of TLR/IL-1R via a homophilic interaction. To date, five TIR domain-containing adaptor molecules have been identi-fied in mammals, including myeloid differentiation factor 88 (MyD88), TIR domain-containing adaptor inducing IFNβ (TRIF), TIR domain-containing adaptor protein (TIRAP)/MyD88 adaptor-like (MAL), TRIF-related adaptor molecule (TRAM) and sterile α and HEAT-Armadillo motifs (SARM). Four of these adaptors (MyD88, TRIF, TIRAP/Mal and TRAM) are known to be involved in TLR signaling in mammals (Figure 2.1). In contrast, a role for mammalian SARM has not yet been identified, although the *Caenorhabditis elegans* SARM homologue has been implicated in host defense against bacteria[19,20]. A distinct set of TIR domain-containing adaptors is recruited to each TLR, and a distinct response is evoked, at least in part, by the combination of adaptors recruited. Among the adaptor molecules, MyD88 and TRIF regulate the activation of the two main signaling pathways leading to the production of proinflammatory cytokines and type I IFNs, respectively. In addition, cell type-specific recruitment of

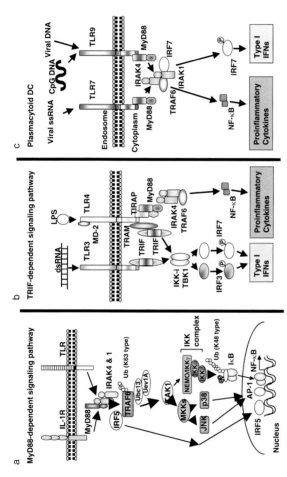

Figure 2.1. Scheme of the TLR-signaling pathways. (a) MyD88-dependent signaling pathways. The signaling pathways emanating from MyD88 activate transcription factors responsible for the induction of proinflammatory cytokine genes. The recruitment of MyD88 to receptors induces sequential activation of IRAKs, TRAF6, TAK1, the IKK complex and NF-κB, which in turn activate MAP kinases such as JNK and p38, thereby inducing AP-1 activation. IRF-5 associates with TRAF6 and MyD88, and translocates to the nucleus in response to stimulation. (b) TRIF-dependent signaling pathways. In cDCs, the TRIF-dependent signaling pathways emanating from TLR3 and TLR4 lead to the induction of both proinflammatory cytokines and type I IFNs. TRIF interacts with TRAF6, RIP1 and TBK1. TRAF6 and RIP1 activate NF-κB, whereas TBK1 and IKK-i phosphorylate IRF-3 and IRF-7, thereby leading to the induction of type I IFN genes. (c) Mechanisms of type I IFN induction in pDCs. TLR7 and TLR9 recruit a complex of MyD88, IRAK1, IRAK4, TRAF6 and IRF-7 in pDCs. Stimulation with CpG-DNA results in the phosphorylation and nuclear translocation of IRF-7, followed by the expression of IFN-α genes. TLR7 and TLR9 stimulation also activates NF-κB through IRAK4 in a TRAF6-dependent manner.

signaling machinery to each receptor is also important for the adaptive response to TLR ligands. In the following sections, we focus on the mechanisms for how TLRs stimulate distinct signaling pathways leading to adaptive responses.

2.3 MyD88-DEPENDENT SIGNALING: A COMMON PATHWAY FOR MOST TLRs FOR THE INDUCTION OF PROINFLAMMATORY CYTOKINES

MyD88 was the first adaptor molecule identified for TLR/IL-1R signaling[21,22]. The MyD88-dependent signaling pathways are shared by all the known TLRs, except for TLR3. MyD88 is comprised of an N-terminal death domain (DD) and a C-terminal TIR domain, and is recruited to the TIR domain of TLR/IL-1R in response to ligand stimulation. Analysis of MyD88-deficient (MyD88$^{-/-}$) mice revealed an essential role for this adaptor in IL-1R/TLR signaling[23–25]. since TLR2-, 5-, 7- and 9-mediated activation of nuclear factor (NF)-κB was abrogated in these mice. Another of the adaptor molecules, TIRAP/Mal, is also involved in MyD88-dependent signaling[26–29]. TIRAP/Mal is specifically responsible for TLR2 and TLR4 signaling, and has been suggested to play a role in bridging TLR2 and TLR4 to MyD88.

Upon recruitment to a TLR, MyD88 interacts with IL-1R-associated kinases (IRAKs), including IRAK1 and IRAK4 (Figure 2.1). IRAKs are composed of an N-terminal DD and an adjacent serine/threonine kinase domain. Interaction of IRAK1 with IRAK4 induces IRAK1 phosphorylation and activation[30]. IL-1R- and TLR-mediated signaling as well as cytokine production were abrogated in IRAK4$^{-/-}$ mice[31]. Although IRAK1$^{-/-}$ mice were reported to show impaired cytokine production in response to IL-1 and LPS stimulation, the mice were still capable of producing significant amounts of cytokines[32,33]. This may be due to compensation by other IRAK family members. The hyperphosphorylated IRAK1 recruits TRAF6 to the receptor complex[34]. TRAF6 contains an N-terminal RING domain and a conserved C-terminal TRAF domain. RING finger domains are also found in the large family of E3 ubiquitin ligases, and TRAF6 was recently shown to function as a ubiquitin ligase, together with an E2 ligase complex consisting of Ubc13 and Uev1A, and catalyze the formation of a lysine-63 (K63)-linked polyubiquitin chain of TRAF6 and NFκB essential modulator (NEMO)/IκB kinase (IKK)γ[35]. Ubiquitinated TRAF6 activates a complex of transforming growth factor (TGF)β-activating kinase 1 (TAK1) and its associated proteins, tubulin antisense-binding protein (TAB)1, TAB2 and TAB3. TAK1 belongs to the mitogen-activated protein kinase kinase kinase (MAPKKK) family, and has been reported to be involved in the IL-1β signaling pathway[36]. Ubiquitin-dependent activation of TAK1 results in the phosphorylation of IKKβ, thereby leading to NF-κB activation. Knockdown of TAK1 by small interfering RNA (siRNA) abolished the IKK activation induced by TNFα and IL-1β stimulation. TAB1, TAB2 and TAB3 were identified as adaptor proteins that associate with TAK1[37,38]. TAB2 and TAB3 bind preferentially to K63-linked polyubiquitin chains through their zinc finger domains. However, IL-1β-mediated activation of NF-κB and MAP kinases was not impaired in mouse embryonic fibroblasts (MEFs) from TAB2$^{-/-}$ mice[39]. Therefore, TAB2 and TAB3 may compensate for each other's functions in vivo. The activated IκB kinase (IKK) complex, composed of IKKα, IKKβ and NEMO/IKKγ, phosphorylates IκB[40]. In resting cells, IκB forms a complex with NF-κB, thereby sequestering NF-κB in the cytoplasm and preventing its activation. Phosphorylated IκB is recognized and modified by K48-linked polyubiquitin by an E3-ubiquitin ligase called bTrCP, resulting in its proteasome-mediated degradation. Following degradation of IκB, the released NF-κB translocates to the nucleus where it binds to promoter regions containing κB sites to induce the expression of proinflammatory cytokine genes.

In addition to NF-κB, MyD88-dependent signaling also activates MAP kinases, including c-Jun N-terminal kinase (JNK) and p38. In turn, the activated MAP kinases phosphorylate and activate AP-1 to induce the expression of proinflammatory cytokine genes in cooperation with NF-κB. The signaling pathway leading to MAP kinase activation is less well understood. TAK1 is capable of phosphorylating MAP kinase kinase (MKK) 6 in IL-1 signaling[36]. JNK and p38 are activated by the upstream MKK3 and MKK6, respectively. A small G protein, Ras, has also been implicated in IL-1-mediated activation of MKK3 and MKK6[41]. Recently, another MAPKKK, apoptosis signal-regulating kinase 1 (ASK1), was reported to be responsible for p38 activation in response to LPS, but not TLR2 or other ligands[42]. ASK1$^{-/-}$ mice showed impaired cytokine responses against LPS. Furthermore, ASK1 can interact with TRAF6, and LPS-induced production of intracellular reactive oxygen species was impaired in ASK1$^{-/-}$ cells.

In addition to NF-κB and MAP kinases, a transcription factor known as IFN-regulatory factor (IRF) 5 is involved in the expression of proinflammatory cytokines. IRF-5 interacts with MyD88 and TRAF6, and ligand stimulation induces its translocation to the nucleus where it activates cytokine gene transcription[43]. DCs from IRF-5$^{-/-}$ mice showed impaired productions of IL-6, TNF-α and IL-12 in response to various TLR stimuli. Although an in vitro study found that IRF-5 was involved in TLR7-mediated type I IFN production[44], it has also been reported to be dispensable for TLR-induced IFN production[43].

TLR signaling is negatively regulated by various molecules, including several transmembrane receptors. T1/ST2 belongs to the IL-1R family and has been suggested as a molecule responsible for Th2 development. In addition, T1/ST2 suppresses IL-1R and TLR4 signaling by sequestering MyD88 and TIRAP/Mal[45]. RP105, a LRR-containing membrane protein, has been implicated in the negative regulation of TLR4 signaling[46]. However, this protein functions as a positive regulator of LPS signaling in B cells[47]. Triad3A, a RING finger protein, functions as a ubiquitin ligase and catalyzes the ubiquitination and degradation of TLR4 and TLR9[48]. Knockdown of Triad3A by siRNA enhanced the responses to TLR ligands. In addition, DAP12, an adaptor protein containing an immunoreceptor tyrosine-based activation motif (ITAM) for receptor signaling on natural killer cells, also negatively regulates TLR signaling[49]. DAP12$^{-/-}$ mice showed increased production of cytokines, enhanced susceptibility to LPS-induced shock and enhanced resistance to infection by *Listeria monocytogenes*[49]. Intracellular signaling molecules, including MyD88, IRAKs and TRAF6, are also modulated and negatively regulated to prevent aberrant activation of inflammatory responses. A splicing variant of MyD88, MyD88(S), lacking the intermediate region, behaves as a dominant-negative regulator of IL-1R and TLR4 signaling[50]. Among the four IRAK family members, IRAK-M has been reported to function as a negative regulator. Tollip, a small protein with an internal C2 domain and a C-terminal CUE domain, has also been implicated in IL-1R/TLR signaling[51]. Tollip associates with IRAK1, and its overexpression perturbs IL-1R-, TLR2- and TLR4-mediated signaling pathways, thereby suggesting that it acts as a negative regulator of TLR signaling[52]. A20, an enzyme that modifies ubiquitination of TRAF6, is required for the termination of TLR-induced NF-κB activation[53].

2.4 TRIF-DEPENDENT PATHWAY: CONNECTING TLR SIGNALING TO TYPE I IFNS

TRIF is another TIR domain-containing adaptor molecule that is responsible for TLR3 and TLR4 signaling (Figure 2.1b)[54,55]. TRIF associates with TLR3 or another of the adaptors, TRAM, via their TIR domains. Since TLR3 and TLR4 can recognize viral components, a feature of TRIF is its ability to induce type I IFNs. Generation of TRIF$^{-/-}$ mice and studies on a mouse strain harboring a mutation in TRIF revealed that TRIF is essential for the responses induced by TLR3 and TLR4 ligands. TRIF$^{-/-}$ mice showed defective production of proinflammatory cytokines and expression of IFN-inducible genes in response to LPS or poly I:C. Furthermore, LPS-mediated activation of IRF-3 was abrogated in TRIF$^{-/-}$ macrophages.

Recent studies have revealed that two IKK-related kinases, namely inducible IκB kinase (IKK-i; also known as IKKε) and TRAF-family member-associated NF-κB activator (TANK)-binding kinase 1 (TBK1; also known as T2K), are involved in IFN production upon TLR stimulation and viral infection[56,57]. IKK-i and TBK1 can directly phosphorylate interferon regulatory factor 3 (IRF-3) and IRF-7 in vitro and activate the IFNβ promoter[56]. TBK1$^{-/-}$, but not IKK-i$^{-/-}$, MEFs showed severely impaired induction of IFNβ and IFN-inducible genes in response to LPS, intracellular introduction of poly I:C and RNA virus infection. Activation of IRF-3, but not NF-κB, in response to LPS and poly(I:C) was also diminished in TBK1$^{-/-}$ cells. IKK-i/TBK1 double-deficient MEFs failed to express any detectable levels of IFNβ and IFN-inducible genes in response to poly I:C, indicating that both IKK-i and TBK1 contribute to the IFN pathway[58–60].

The transcription factors IRF-3 and IRF-7 are essential for the expression of type I IFNs in response to viral infection and TLR stimulation. Following phosphorylation by TBK1/IKK-i, IRF-3 and IRF-7 form homodimers or heterodimers and translocate into the nucleus. Once there, they interact with the coactivator proteins CBP and p300 and subsequently bind to IFN-stimulated regulatory elements (ISREs) present in the promoters of a set of IFN-inducible genes to induce their expression[61]. Secreted type I IFNs, in turn, activate their receptors in both autocrine and paracrine manners. This activates IFN-stimulated gene factor 3 (ISGF3), which consists of signal transducer and activator of transcription (STAT) 1, STAT2 and p40/IRF-9, thereby amplifying the response. IRF-3$^{-/-}$ mice showed defective induction of IFNβ in response to LPS, indicating a critical role for IRF-3 in the TLR4-induced IFN response[62]. IRF-7$^{-/-}$ mice showed profound defects in the induction of type I IFNs, but not in the activation of NF-κB, in response to CpG-DNA and viral infection, and therefore succumbed to infection with several viruses[63]. Furthermore, cells from IRF-3/IRF-7 double-deficient mice showed completely abrogated production of type I IFNs upon exposure to several viruses, suggesting that both IRF-3 and IRF-7 are involved in the initial production of IFNs in response to viral exposure[63].

TRIF-dependent signaling also activates NF-κB. TRIF interacts with TRAF6 through TRAF6 binding motifs in the N-terminal portion of TRIF, and with receptor interacting protein (RIP1 and RIP3) via a C-terminal RIP homotypic interaction motif (RHIM)[64,65]. Poly I:C-induced NF-κB activation was abolished in RIP1[−/−] MEFs. In addition, overexpression of TRIF was found to induce apoptosis in some cells[66]. Deletion of the C-terminal RHIM domain abolished the activity[67].

TLR4-mediated NF-κB activation is regulated by either MyD88 or TRIF. In TLR4-signaling, MyD88 regulates rapid activation of NF-κB, while TRIF is responsible for sustained activation. Mice lacking both MyD88 and TRIF were unable to activate NF-κB or MAP kinases as well as IRFs, indicating that these two adaptors are responsible for TLR4 signaling[68]. Microarray analyses of LPS-inducible genes in MyD88[−/−], TRIF[−/−] and MyD88[−/−] TRIF[−/−] macrophages revealed that these two adaptors are essential for the induction of all genes in response to LPS stimulation[69]. On the other hand, cells lacking either MyD88 or TRIF could still induce various genes, with the exception of proinflammatory cytokines and IFN-inducible genes. These findings suggest that many NF-κB target genes are ambiguously regulated by MyD88 or TRIF whose signaling governs early or late activation of NF-κB[69].

TRAM, another TIR domain-containing adaptor, is specifically involved in TLR4, but not TLR3, signaling[70,71]. Overexpression of TRAM activates IRF-3 and NF-κB-dependent gene induction. LPS-induced cytokine production and IFN-inducible gene expression were severely impaired in TRAM[−/−] cells. Furthermore, TRAM[−/−] cells failed to sustain LPS-induced NF-κB activation compared to wild-type cells[71]. In contrast, there were no defects in the responses to poly I:C and CpG-DNA. TRAM can interact physically with both TLR4 and TRIF, but no direct association was detected between TLR4 and TRIF. Taken together, these observations suggest that TRAM functions to bridge TLR4 and TRIF, resulting in activation of the downstream signaling pathway.

2.5 TLR7, 9 SIGNALING IN PLASMACYTOID DCs INDUCES PRODUCTION OF TYPE I IFNs

DCs are subcategorized into conventional and plasmacytoid DCs by their expressions of different surface molecules as well as their abilities to produce cytokines and type I IFNs upon exposure to pathogens. Conventional DCs (cDCs) express various TLRs, including TLR3, and stimulation with TLR ligands results in the production of proinflammatory cytokines. Stimulation of cDCs with poly I:C or LPS induces the expression of IFNβ and IFN-inducible genes, indicating that the TRIF-dependent pathway mainly functions to induce type I IFNs in response to TLR stimulation.

A subset of DCs, plasmacytoid DCs (pDCs), expresses high levels of TLR7 and TLR9, but not TLR3. pDCs produce large amounts of IFNα upon exposure to viruses[72]. Ligand stimulation of TLR7 and TLR9 results in the production of IFNα. Among the TLR9 ligands, A/D-type CpG-DNAs, but not conventional B/K-type CpG-DNAs, are potent inducers of IFNα, although the production of proinflammatory cytokines is vigorously induced by both types of CpG-DNAs[73]. In addition, infection with several DNA and RNA viruses also induces IFN-α production by pDCs in a TLR7- or TLR9-dependent manner[14,16,74]. Interestingly, the type I IFN induction in pDCs is independent of TBK1. In pDCs, MyD88 was found to directly associate with IRF-7, which then undergoes modifications such as phosphorylation and ubiquitination (Figure 2.1c)[75,76]. IRAK4, IRAK1 and TRAF6 are also recruited to the complex. MyD88$^{-/-}$ and IRAK4$^{-/-}$ pDCs failed to activate both NF-κB and IRF-7 in response to TLR9 ligands. In contrast, IRAK-1 plays a key role in the expression of type I IFNs, but not proinflammatory cytokines[77]. In IRAK1$^{-/-}$ pDCs, TLR9-mediated nuclear translocation of IRF-7 was specifically abrogated. In addition, IRAK1 can phosphorylate IRF-7 in vitro, suggesting that IRAK1 possibly functions as an IRF-7 kinase. The activated IRF-7 induces the expression of IFNα genes.

The mechanisms of the cell type-specific IFNα production in response to A/D-type CpG-DNAs are intriguing to explore. A/D-type CpG-DNAs are structurally different from conventional CpG-DNAs, since they possess a phosphorothioate-modified poly G stretch at the 5′ and 3′ ends and a phosphodiester CpG motif in the central portion. It has been hypothesized that pDCs, but not cDCs, constitutively express IRF-7, thereby explaining the induction mechanism for type I IFNs[78]. Recently, it was proposed that A/D-type CpG-DNAs are retained in endosomal vesicles in pDCs for a long time, thus facilitating encounters between the DNA and TLR9-MyD88-IRF-7 complexes[79]. In cDCs, A/D-type CpG-DNAs are rapidly moved to lysosomal vesicles. B/K-type CpG-DNAs, which induce lower levels of IFNα production, acquire this activity after treatment with a cationic lipid that maintains the DNA in endosomes in pDCs. However, this study is controversial to several previous findings. First, although A/D-type CpG-DNA-induced IFNα production is specific to pDCs, the production of proinflammatory cytokines, including TNF-α, IL-6 and IL-12, is vigorously induced in response to both A/D-type and B/K-type CpG-DNAs in cDCs. These proinflammatory cytokines are also induced in a TLR9-MyD88-dependent fashion, suggesting that MyD88-dependent signaling is not impaired in cDCs. Moreover, low concentrations of B/K-type CpG-DNAs are known to induce IFNα in pDCs, and this IFNα-inducing activity is lost at higher concentrations[73].

These findings imply that the differences between A/D-type and B/K-type DNAs, and between pDCs and cDCs, cannot be fully explained by endosomal retention of CpG-DNAs alone. Future studies are therefore necessary to elucidate the additional mechanisms of the type I IFN production in response to TLR7 and TLR9 ligands in pDCs.

2.6 INTRACELLULAR RECOGNITION OF PATHOGENS

TLRs recognize pathogens at either the cell surface or lysosome/endosome membranes, indicating that this system is invalid for the detection of pathogens that have invaded the cytosol. Recent studies have revealed that such pathogens already present in the cytoplasm are detected by various cytoplasmic receptors without a transmembrane domain.

Synthesis of dsRNA in the infected cells is required for the replication of ssRNA viruses, and such viral dsRNA is believed to be a virus-specific molecular structure recognized by hosts. Most virus-infected cells produce type I IFNs in a TLR3-independent manner, implying that the mechanism for dsRNA recognition in the cytoplasm plays an important role in the IFN response. Protein kinase R (PKR) has been implicated in viral dsRNA recognition. PKR belongs to a family of proteins containing dsRNA-binding domains, and is induced in response to stimulation with IFNs. Activation of PKR results in growth arrest of the cells via eIF2α phosphorylation. Several reports have demonstrated that virus-induced type I IFN production is modestly impaired in PKR$^{-/-}$ mice[80]. Furthermore, PKR$^{-/-}$ mice showed defective induction of apoptosis in response to poly I:C and bacterial burdens. However, PKR$^{-/-}$ mice have also been reported to show no alterations in their IFN responses to viral exposure[81].

Recently, retinoic acid-inducible protein-I (RIG-I), an IFN-inducible protein containing caspase recruitment domains (CARDs) and a helicase domain, has been identified as a cytoplasmic dsRNA detector. RIG-I also interacts with poly I:C[82]. RIG-I overexpression conferred Newcastle Disease virus (NDV)- and dsRNA-mediated IFN responses on the cells. Mda-5, a molecule showing homology to RIG-I, has also been implicated in the recognition of viral dsRNA[83]. RIG-I$^{-/-}$ mice generally showed embryonic lethality due to liver degeneration. RIG-I$^{-/-}$ MEFs did not produce IFNβ, activate IFN-inducible genes or induce proinflammatory cytokines in response to infections with NDV, Sendai virus and Vesicular Stomatitis virus (VSV)[84]. NDV-induced activation of NF-κB and ISRE-containing genes was abrogated in RIG-I$^{-/-}$ cells. RIG-I overexpression in TBK1/IKK-i double-deficient cells failed to activate the IFNβ promoter, indicating that

RIG-I acts upstream of TBK1/IKK-i and governs both NF-κB and IRFs. Furthermore, cDCs derived from the few RIG-I$^{-/-}$ mice born alive also failed to induce type I IFNs and cytokines upon exposure to NDV. Interestingly, pDCs from RIG-I$^{-/-}$ mice produced a comparable amount of IFNα to pDCs from wild-type mice upon NDV stimulation, indicating that cDCs and pDCs mainly utilize different mechanisms for the induction of IFNs[84].

In MEFs, FADD and RIP1 are required for type I IFN responses to stimulation with dsRNA, and FADD$^{-/-}$ or RIP1$^{-/-}$ MEFs were highly susceptible to VSV infection[85]. FADD and RIP1 act upstream of TBK1/IKK-i, suggesting that these molecules transduce signals from RIG-I to TBK1. However, the cited report[85] only described the response to poly I:C, and not that against viral infection. Further analyses are therefore required to clarify the roles of FADD and RIP1 in the IFN response to viral infection. The mechanism of RIG-I-induced signaling as well as the relationships among RIG-I, Mda-5 and TLR signaling will be extremely interesting to explore.

CARD-containing proteins are also involved in recognizing bacteria in the cytoplasm. These proteins include nucleotide-binding oligomerization domain (NOD1 and NOD2), which both contain N-terminal CARD domains. NOD1 detects γ-d-glutamyl-meso diaminopimelic acid (iE-DAP) found in Gram-negative bacterial peptidoglycan[86,87], and NOD1$^{-/-}$ macrophages failed to produce cytokines in response to iE-DAP[87]. In contrast, NOD2 is a receptor for muramyl-dipeptide (MDP) derived from bacterial peptidoglycan (PGN)[88]. A missense point mutation in the human *Nod2* gene is correlated with susceptibility to Crohn's disease, an inflammatory bowel disease. Overexpression of a mutant NOD2 in cells failed to induce NF-κB activation[89]. In addition, NOD2$^{-/-}$ mice did not respond to MDP. However, the mechanism for the development of this inflammatory disease is still controversial. NOD2$^{-/-}$ mice were reported to show defective induction of intestinal anti-microbial peptides after oral bacterial infection[90], and were susceptible to mucosal bacterial infection, thereby explaining the role of NOD2 in the development of Crohn's disease[90]. Another study reported that NOD2 suppressed TLR2-mediated activation of NF-κB in response to PGN[91]. NOD2$^{-/-}$ mice showed enhanced IL-12 production in response to PGN stimulation, which may explain the mechanism for how NOD2 controls the susceptibility to this disease[91]. Furthermore, mice with a mutation corresponding to that of Crohn's disease showed enhanced NF-κB activation in response to MDP, and increased susceptibility to dextran sodium sulfate (DSS)-induced colitis[92]. The mutant mice exhibited elevated NF-κB activation in response to MDP. Given the controversy between these separate studies, future studies are required to clarify the mechanism for the function of NOD2 in intestinal immune cells. Ligand binding to NOD1 and NOD2 causes their oligomerization and results in NF-κB activation through the recruitment of RIP2/Rip-like interacting caspase-like apoptosis-regulatory protein kinase (RICK), a serine/threonine kinase[93]. NODs associate with RIP2/RICK through their CARD domains by homophilic interactions. It has been revealed that proteins with a nucleotide-binding oligomerization domain (NOD) and a LRR domain form a large group, called the NOD-LRR family[94]. The members of this family share a tripartite domain structure consisting of C-terminal LRR motifs, a central NOD and N-terminal protein–protein interaction motifs, such as CARDs, pyrin domains or a TIR domain[94]. Although the number of NOD-LRR family members is continuing to increase, the functions of members other than NOD1 and NOD2 have not yet been clarified.

2.7 DISTURBANCE OF TLR SIGNALING BY PATHOGENS

Activation of the immune system is a threat to the survival of pathogens. Therefore, viruses and bacteria have developed elaborate mechanisms to escape from host immune surveillance. Pathogens modify and/or inhibit host immune responses, such as the induction of IFNs and proinflammatory cytokines, and inhibit apoptosis.

The TLR system is a target of viral immune disturbance. Vaccinia virus encodes the TIR domain-containing proteins A46R and A52R[95,96]. A46R targets host TIR domain-containing adaptors and suppresses TLR- or IL-1R-induced NF-κB activation. Similarly, A52R suppresses TLR-mediated signaling. However, A46R and A52R each have distinct roles in TLR3-mediated signaling: A46R inhibits IRF-3, whereas A52R blocks NF-κB[96]. Vaccinia virus also encodes another protein, N1L, that antagonizes TLR signaling at the level of IκB kinases and TBK1[97].

Hepatitis C virus (HCV) also has a mechanism for suppressing TLR signaling. HCV NS3/4A protease was shown to block IRF-3 activation by cleaving TRIF[98]. This cleavage decreases the abundance of TRIF and inhibits poly I:C-mediated activation of IFN responses. The RIG-I system is considered to be another viral target for immune evasion, and NS3/4A was also able to inhibit RIG-I mediated responses in a TRIF-independent manner[99]. The V proteins of paramyxoviruses associate with Mda-5, and inhibit dsRNA-induced activation of the IFNβ promoter[83].

Bacteria and fungi exploit the TLR system to evade host immune responses. For instance, *Mycobacterium tuberculosis* avoids being killed by macrophages by inhibiting IFN-γ-mediated signaling. Prolonged signaling with a 19-kDa lipoprotein from *Mycobacterium*, which stimulates TLR2, inhibits IFNγ production and major histocompatibility complex (MHC) class II antigen processing activity[100]. These findings suggest that, at least in part, TLR2 signaling promotes evasion of T cell responses and the persistence of *Mycobacterium*. *Candida albicans* was also reported to exploit TLR2 signaling to suppress immunity. TLR2$^{-/-}$ mice were more resistant to systemic *Candida* infection and IL-10 release was impaired in TLR2$^{-/-}$ mice[101].

These observations indicate that the TLR signaling pathways and additional signaling pathways involved in pathogen recognition are inhibited and exploited by pathogens for their survival.

2.8 PERSPECTIVES

In this chapter, we have focused on the signaling pathways activated in response to exposure to pathogens. Although all TLRs possess a conserved TIR domain that is critical for the activation of intracellular signaling, each TLR induces a distinct response. This can partly be explained by the recruitment of different TIR domain-containing adaptor molecules to each receptor. In addition, cell type-specific recruitment of distinct signaling machinery is also responsible for selective IFN-α production.

Recent extensive studies have identified ligands for most of the TLRs as well as their signaling pathways. TLR2 and TLR5, which recognize bacterial, but not viral, components, are completely dependent on MyD88 for intracellular signaling. Therefore, the production of proinflammatory cytokines, but not IFNs, is induced in response to TLR2 and TLR5 ligands. In contrast, TLR3 and TLR7 recognize virus-specific components, while TLR4 and TLR9 can respond to both bacteria and viruses. TLRs that recognize viral components induce the production of type I IFNs in response to proinflammatory cytokines. This adaptive response depends on the type of pathogen and can be explained by the adaptor molecules recruited to each TLR or by a cell type-specific mechanism.

Recent studies have rapidly identified the TLR signaling pathways activated by distinct TLR ligands. In addition, intracellular receptors for pathogens and their signaling pathways are beginning to be identified. However, the complete mechanisms involved in immune-recognition and signaling are far from comprehensively understood. In addition, immune cells utilize different signaling pathways in a cell type-specific manner. Future studies are necessary to clarify the mechanisms and develop techniques for manipulating the pathways to improve therapeutic treatments for infectious diseases.

REFERENCES

1. Akira, S. and K. Takeda (2004). Toll-like receptor signalling. *Nat. Rev. Immunol.* **4**: 499–511.
2. Beutler, B. (2004). Inferences, questions and possibilities in Toll-like receptor signalling. *Nature* **430**: 257–63.
3. Poltorak, A., X. He, I. Smirnova, M. Y. Liu, C. Van Huffel, X. Du, D. Birdwell, E. Alejos, M. Silva, C. Galanos, M. Freudenberg, P. Ricciardi-Castagnoli, B. Layton, and B. Beutler (1998). Defective LPS signaling in C3H/HeJ and C57BL/10ScCr mice: mutations in Tlr4 gene. *Science* **282**: 2085–8.
4. Hoshino, K., O. Takeuchi, T. Kawai, H. Sanjo, T. Ogawa, Y. Takeda, K. Takeda, and S. Akira (1999). Cutting edge: Toll-like receptor 4 (TLR4)-deficient mice are hyporesponsive to lipopolysaccharide: evidence for TLR4 as the Lps gene product. *J. Immunol.* **162**: 3749–52.
5. Takeuchi, O., K. Hoshino, T. Kawai, H. Sanjo, H. Takada, T. Ogawa, K. Takeda, and S. Akira (1999). Differential roles of TLR2 and TLR4 in recognition of Gram-negative and Gram-positive bacterial cell wall components. *Immunity* **11**: 443–51.
6. Takeuchi, O., T. Kawai, P. F. Muhlradt, M. Morr, J. D. Radolf, A. Zychlinsky, K. Takeda, and S. Akira (2001). Discrimination of bacterial lipoproteins by Toll-like receptor 6. *Int. Immunol.* **13**: 933–40.
7. Takeuchi, O., S. Sato, T. Horiuchi, K. Hoshino, K. Takeda, Z. Dong, R. L. Modlin, and S. Akira (2002). Cutting edge: role of Toll-like receptor 1 in mediating immune response to microbial lipoproteins. *J. Immunol.* **169**: 10–14.
8. Alexopoulou, L., V. Thomas, M. Schnare, Y. Lobet, J. Anguita, R. T. Schoen, R. Medzhitov, E. Fikrig, and R. A. Flavell (2002). Hyporesponsiveness to vaccination with *Borrelia burgdorferi* OspA in humans and in TLR1- and TLR2-deficient mice. *Nat. Med.* **8**: 878–84.
9. Hayashi, F., K. D. Smith, A. Ozinsky, T. R. Hawn, E. C. Yi, D. R. Goodlett, J. K. Eng, S. Akira, D. M. Underhill, and A. Aderem (2001). The innate immune response to bacterial flagellin is mediated by Toll-like receptor 5. *Nature* **410**: 1099–103.

10. Zhang, D., G. Zhang, M. S. Hayden, M. B. Greenblatt, C. Bussey, R. A. Flavell, and S. Ghosh (2004). A toll-like receptor that prevents infection by uropathogenic bacteria. *Science* **303**: 1522–6.

11. Yarovinsky, F., D. Zhang, J. F. Andersen, G. L. Bannenberg, C. N. Serhan, M. S. Hayden, S. Hieny, F. S. Sutterwala, R. A. Flavell, S. Ghosh, and A. Sher (2005). TLR11 Activation of dendritic cells by a protozoan profilin-like protein. *Science* **308**: 1626–9.

12. Alexopoulou, L., A. C. Holt, R. Medzhitov, and R. A. Flavell (2001). Recognition of double-stranded RNA and activation of NF-kappaB by Toll-like receptor 3. *Nature* **413**: 732–8.

13. Heil, F., H. Hemmi, H. Hochrein, F. Ampenberger, C. Kirschning, S. Akira, G. Lipford, H. Wagner, and S. Bauer (2004). Species-specific recognition of single-stranded RNA via Toll-like receptor 7 and 8. *Science* **303**: 1526–9.

14. Diebold, S. S., T. Kaisho, H. Hemmi, S. Akira, and C. Reis e Sousa (2004). Innate antiviral responses by means of TLR7-mediated recognition of single-stranded RNA. *Science* **303**: 1529–31.

15. Hemmi, H., O. Takeuchi, T. Kawai, T. Kaisho, S. Sato, H. Sanjo, M. Matsumoto, K. Hoshino, H. Wagner, K. Takeda, and S. Akira (2000). A Toll-like receptor recognizes bacterial DNA. *Nature* **408**: 740–5.

16. Lund, J., A. Sato, S. Akira, R. Medzhitov, and A. Iwasaki (2003). Toll-like receptor 9-mediated recognition of Herpes simplex virus-2 by plasmacytoid dendritic cells. *J. Exp. Med.* **198**: 513–20.

17. Krug, A., G. D. Luker, W. Barchet, D. A. Leib, S. Akira, and M. Colonna (2004). Herpes simplex virus type 1 activates murine natural interferon-producing cells through Toll-like receptor 9. *Blood* **103**: 1433–7.

18. Latz, E., A. Schoenemeyer, A. Visintin, K. A. Fitzgerald, B. G. Monks, C. F. Knetter, E. Lien, N. J. Nilsen, T. Espevik, and D. T. Golenbock (2004). TLR9 signals after translocating from the ER to CpG DNA in the lysosome. *Nat. Immunol.* **5**: 190–8.

19. Couillault, C., N. Pujol, J. Reboul, L. Sabatier, J. F. Guichou, Y. Kohara, and J. J. Ewbank (2004). TLR-independent control of innate immunity in *Caenorhabditis elegans* by the TIR domain adaptor protein TIR-1, an ortholog of human SARM. *Nat. Immunol.* **5**: 488–94.

20. Liberati, N. T., K. A. Fitzgerald, D. H. Kim, R. Feinbaum, D. T. Golenbock, and F. M. Ausubel (2004). Requirement for a conserved Toll/interleukin-1 resistance domain protein in the *Caenorhabditis elegans* immune response. *Proc. Natl Acad. Sci. U S A* **101**: 6593–8.

21. Muzio, M., J. Ni, P. Feng, and V. M. Dixit (1997). IRAK (Pelle) family member IRAK-2 and MyD88 as proximal mediators of IL-1 signaling. *Science* **278**: 1612–15.

22. Wesche, H., W. J. Henzel, W. Shillinglaw, S. Li, and Z. Cao (1997). MyD88: an adapter that recruits IRAK to the IL-1 receptor complex. *Immunity* **7**: 837–47.

23. Adachi, O., T. Kawai, K. Takeda, M. Matsumoto, H. Tsutsui, M. Sakagami, K. Nakanishi, and S. Akira (1998). Targeted disruption of the MyD88 gene results in loss of IL-1- and IL-18-mediated function. *Immunity* **9**: 143–50.

24. Kawai, T., O. Adachi, T. Ogawa, K. Takeda, and S. Akira (1999). Unresponsiveness of MyD88-deficient mice to endotoxin. *Immunity* **11**: 115–22.

25. Takeuchi, O., K. Takeda, K. Hoshino, O. Adachi, T. Ogawa, and S. Akira (2000). Cellular responses to bacterial cell wall components are mediated through MyD88-dependent signaling cascades. *Int. Immunol.* **12**: 113–17.

26. Horng, T., G. M. Barton, and R. Medzhitov (2001). TIRAP: an adapter molecule in the Toll signaling pathway. *Nat. Immunol.* **2**: 835–41.

27. Fitzgerald, K. A., E. M. Palsson-McDermott, A. G. Bowie, C. A. Jefferies, A. S. Mansell, G. Brady, E. Brint, A. Dunne, P. Gray, M. T. Harte, D. McMurray, D. E. Smith, J. E. Sims, T. A. Bird, and L. A. O'Neill (2001). Mal (MyD88-adapter-like) is required for Toll-like receptor-4 signal transduction. *Nature* **413**: 78–83.

28. Horng, T., G. M. Barton, R. A. Flavell, and R. Medzhitov (2002). The adaptor molecule TIRAP provides signalling specificity for Toll-like receptors. *Nature* **420**: 329–33.

29. Yamamoto, M., S. Sato, H. Hemmi, H. Sanjo, S. Uematsu, T. Kaisho, K. Hoshino, O. Takeuchi, M. Kobayashi, T. Fujita, K. Takeda, and S. Akira (2002). Essential role for TIRAP in activation of the signalling cascade shared by TLR2 and TLR4. *Nature* **420**: 324–9.

30. Li, S., A. Strelow, E. J. Fontana, and H. Wesche (2002). IRAK-4: a novel member of the IRAK family with the properties of an IRAK-kinase. *Proc. Natl Acad. Sci. U S A* **99**: 5567–72.

31. Suzuki, N., S. Suzuki, G. S. Duncan, D. G. Millar, T. Wada, C. Mirtsos, H. Takada, A. Wakeham, A. Itie, S. Li, J. M. Penninger, H. Wesche, P. S. Ohashi, T. W. Mak, and W. C. Yeh (2002). Severe impairment of interleukin-1 and Toll-like receptor signalling in mice lacking IRAK-4. *Nature* **416**: 750–6.

32. Thomas, J. A., J. L. Allen, M. Tsen, T. Dubnicoff, J. Danao, X. C. Liao, Z. Cao, and S. A. Wasserman (1999). Impaired cytokine signaling in mice lacking the IL-1 receptor-associated kinase. *J. Immunol.* **163**: 978–84.

33. Swantek, J. L., M. F. Tsen, M. H. Cobb, and J. A. Thomas (2000). IL-1 receptor-associated kinase modulates host responsiveness to endotoxin. *J. Immunol.* **164**: 4301–6.

34. Jiang, Z., J. Ninomiya-Tsuji, Y. Qian, K. Matsumoto, and X. Li (2002). Interleukin-1 (IL-1) receptor-associated kinase-dependent IL-1-induced signaling complexes phosphorylate TAK1 and TAB2 at the plasma membrane and activate TAK1 in the cytosol. *Mol. Cell Biol.* **22**: 7158–67.

35. Deng, L., C. Wang, E. Spencer, L. Yang, A. Braun, J. You, C. Slaughter, C. Pickart, and Z. J. Chen (2000). Activation of the IkappaB kinase complex by TRAF6 requires a dimeric ubiquitin-conjugating enzyme complex and a unique polyubiquitin chain. *Cell* **103**: 351–61.

36. Wang, C., L. Deng, M. Hong, G. R. Akkaraju, J. Inoue, and Z. J. Chen (2001). TAK1 is a ubiquitin-dependent kinase of MKK and IKK. *Nature* **412**: 346–51.

37. Ishitani, T., G. Takaesu, J. Ninomiya-Tsuji, H. Shibuya, R. B. Gaynor, and K. Matsumoto (2003). Role of the TAB2-related protein TAB3 in IL-1 and TNF signaling. *EMBO J.* **22**: 6277–88.

38. Kanayama, A., R. B. Seth, L. Sun, C. K. Ea, M. Hong, A. Shaito, Y. H. Chiu, L. Deng, and Z. J. Chen (2004). TAB2 and TAB3 activate the NF-kappaB pathway through binding to polyubiquitin chains. *Mol. Cell* **15**: 535–48.

39. Sanjo, H., K. Takeda, T. Tsujimura, J. Ninomiya-Tsuji, K. Matsumoto, and S. Akira (2003). TAB2 is essential for prevention of apoptosis in fetal liver but not for interleukin-1 signaling. *Mol. Cell Biol.* **23**: 1231–8.

40. Hayden, M. S. and S. Ghosh (2004). Signaling to NF-kappaB. *Genes Dev.* **18**: 2195–224.

41. McDermott, E. P. and L. A. O'Neill (2002). Ras participates in the activation of p38 MAPK by interleukin-1 by associating with IRAK, IRAK2, TRAF6, and TAK-1. *J. Biol. Chem.* **277**: 7808–15.

42. Matsuzawa, A., K. Saegusa, T. Noguchi, C. Sadamitsu, H. Nishitoh, S. Nagai, S. Koyasu, K. Matsumoto, K. Takeda, and H. Ichijo (2005). ROS-dependent activation of the TRAF6-ASK1-p38 pathway is selectively required for TLR4-mediated innate immunity. *Nat. Immunol.* **6**: 587–92.

43. Takaoka, A., H. Yanai, S. Kondo, G. Duncan, H. Negishi, T. Mizutani, S. Kano, K. Honda, Y. Ohba, T. W. Mak, and T. Taniguchi (2005). Integral role of IRF-5 in the gene induction programme activated by Toll-like receptors. *Nature* **434**: 243–9.

44. Schoenemeyer, A., B. J. Barnes, M. E. Mancl, E. Latz, N. Goutagny, P. M. Pitha, K. A. Fitzgerald, and D. T. Golenbock (2005). The interferon regulatory factor, IRF5, is a central mediator of toll-like receptor 7 signaling. *J. Biol. Chem.* **280**: 17005–12.

45. Brint, E. K., D. Xu, H. Liu, A. Dunne, A. N. McKenzie, L. A. O'Neill, and F. Y. Liew (2004). ST2 is an inhibitor of interleukin 1 receptor and Toll-like receptor 4 signaling and maintains endotoxin tolerance. *Nat. Immunol.* **5**: 373–9.

46. Divanovic, S., A. Trompette, S. F. Atabani, R. Madan, D. T. Golenbock, A. Visintin, R. W. Finberg, A. Tarakhovsky, S. N. Vogel, Y. Belkaid, E. A. Kurt-Jones, and C. L. Karp (2005). Negative regulation of Toll-like receptor 4 signaling by the Toll-like receptor homolog RP105. *Nat. Immunol.* **6**: 571–8.

47. Nagai, Y., T. Kobayashi, Y. Motoi, K. Ishiguro, S. Akashi, S. Saitoh, Y. Kusumoto, T. Kaisho, S. Akira, M. Matsumoto, K. Takatsu, and K. Miyake (2005). The radioprotective 105/MD-1 complex links TLR2 and TLR4/MD-2 in antibody response to microbial membranes. *J. Immunol.* **174**: 7043–9.

48. Chuang, T. H. and R. J. Ulevitch (2004). Triad3A, an E3 ubiquitin-protein ligase regulating Toll-like receptors. *Nat. Immunol.* **5**: 495–502.

49. Hamerman, J. A., N. K. Tchao, C. A. Lowell, and L. L. Lanier (2005). Enhanced Toll-like receptor responses in the absence of signaling adaptor DAP12. *Nat. Immunol.* **6**: 579–86.

50. Burns, K., S. Janssens, B. Brissoni, N. Olivos, R. Beyaert, and J. Tschopp (2003). Inhibition of interleukin 1 receptor/Toll-like receptor signaling through the alternatively spliced, short form of MyD88 is due to its failure to recruit IRAK-4. *J. Exp. Med.* **197**: 263–8.

51. Burns, K., J. Clatworthy, L. Martin, F. Martinon, C. Plumpton, B. Maschera, A. Lewis, K. Ray, J. Tschopp, and F. Volpe (2000). Tollip, a new component of the IL-1RI pathway, links IRAK to the IL-1 receptor. *Nat. Cell Biol.* **2**: 346–51.

52. Zhang, G. and S. Ghosh (2002). Negative regulation of Toll-like receptor-mediated signaling by Tollip. *J. Biol. Chem.* **277**: 7059–65.

53. Boone, D. L., E. E. Turer, E. G. Lee, R. C. Ahmad, M. T. Wheeler, C. Tsui, P. Hurley, M. Chien, S. Chai, O. Hitotsumatsu, E. McNally, C. Pickart, and A. Ma (2004). The ubiquitin-modifying enzyme A20 is required for termination of Toll-like receptor responses. *Nat. Immunol.* **5**: 1052–60.

54. Yamamoto, M., S. Sato, K. Mori, K. Hoshino, O. Takeuchi, K. Takeda, and S. Akira (2002). Cutting edge: a novel Toll/IL-1 receptor domain-containing adapter that preferentially activates the IFN-beta promoter in the Toll-like receptor signaling. *J. Immunol.* **169**: 6668–72.

55. Oshiumi, H., M. Matsumoto, K. Funami, T. Akazawa, and T. Seya (2003). TICAM-1, an adaptor molecule that participates in Toll-like receptor 3-mediated interferon-beta induction. *Nat. Immunol.* **4**: 161–7.

56. Sharma, S., B. R. tenOever, N. Grandvaux, G. P. Zhou, R. Lin, and J. Hiscott (2003). Triggering the interferon antiviral response through an IKK-related pathway. *Science* **300**: 1148–51.

57. Fitzgerald, K. A., S. M. McWhirter, K. L. Faia, D. C. Rowe, E. Latz, D. T. Golenbock, A. J. Coyle, S. M. Liao, and T. Maniatis (2003). IKKepsilon and TBK1 are essential components of the IRF3 signaling pathway. *Nat. Immunol.* **4**: 491–6.

58. Hemmi, H., O. Takeuchi, S. Sato, M. Yamamoto, T. Kaisho, H. Sanjo, T. Kawai, K. Hoshino, K. Takeda, and S. Akira (2004). The roles of two IkappaB kinase-related kinases in lipopolysaccharide and double stranded RNA signaling and viral infection. *J. Exp. Med.* **199**: 1641–50.

59. McWhirter, S. M., K. A. Fitzgerald, J. Rosains, D. C. Rowe, D. T. Golenbock, and T. Maniatis (2004). IFN-regulatory factor 3-dependent gene expression is defective in Tbk1-deficient mouse embryonic fibroblasts. *Proc. Natl Acad. Sci. U S A* **101**: 233–8.

60. Perry, A. K., E. K. Chow, J. B. Goodnough, W. C. Yeh, and G. Cheng (2004). Differential requirement for TANK-binding kinase-1 in type I interferon responses to Toll-like receptor activation and viral infection. *J. Exp. Med.* **199**: 1651–8.

61. Yoneyama, M., W. Suhara, Y. Fukuhara, M. Fukuda, E. Nishida, and T. Fujita (1998). Direct triggering of the type I interferon system by virus infection: activation of a transcription factor complex containing IRF-3 and CBP/p300. *EMBO J.* **17**: 1087–95.

62. Sakaguchi, S., H. Negishi, M. Asagiri, C. Nakajima, T. Mizutani, A. Takaoka, K. Honda, and T. Taniguchi (2003). Essential role of IRF-3 in lipopolysaccharide-induced interferon-beta gene expression and endotoxin shock. *Biochem. Biophys. Res. Commun.* **306**: 860–6.

63. Honda, K., H. Yanai, H. Negishi, M. Asagiri, M. Sato, T. Mizutani, N. Shimada, Y. Ohba, A. Takaoka, N. Yoshida, and T. Taniguchi (2005). IRF-7 is the master regulator of type-I interferon-dependent immune responses. *Nature* **434**: 772–7.

64. Sato, S., M. Sugiyama, M. Yamamoto, Y. Watanabe, T. Kawai, K. Takeda, and S. Akira (2003). Toll/IL-1 receptor domain-containing adaptor inducing IFN-beta (TRIF) associates with TNF receptor-associated factor 6 and TANK-binding kinase 1, and activates two distinct transcription factors, NF-kappa B and IFN-regulatory factor-3, in the Toll-like receptor signaling. *J. Immunol.* **171**: 4304–10.

65. Meylan, E., K. Burns, K. Hofmann, V. Blancheteau, F. Martinon, M. Kelliher, and J. Tschopp (2004). RIP1 is an essential mediator of Toll-like receptor 3-induced NF-kappaB activation. *Nat. Immunol.* **5**: 503–7.

66. Han, K. J., X. Su, L. G. Xu, L. H. Bin, J. Zhang, and H. B. Shu (2004). Mechanisms of the TRIF-induced interferon-stimulated response element and NF-kappaB activation and apoptosis pathways. *J. Biol. Chem.* **279**: 15652–61.

67. Kaiser, W. J. and M. K. Offermann (2005). Apoptosis induced by the toll-like receptor adaptor TRIF is dependent on its receptor interacting protein homotypic interaction motif. *J. Immunol.* **174**: 4942–52.

68. Yamamoto, M., S. Sato, H. Hemmi, K. Hoshino, T. Kaisho, H. Sanjo, O. Takeuchi, M. Sugiyama, M. Okabe, K. Takeda, and S. Akira (2003). Role of adaptor TRIF in the MyD88-independent toll-like receptor signaling pathway. *Science* **301**: 640–3.

69. Hirotani, T., M. Yamamoto, Y. Kumagai, S. Uematsu, I. Kawase, O. Takeuchi, and S. Akira (2005). Regulation of lipopolysaccharide-inducible genes by MyD88 and Toll/IL-1 domain containing adaptor inducing IFN-beta. *Biochem. Biophys. Res. Commun.* **328**: 383–92.

70. Fitzgerald, K. A., D. C. Rowe, B. J. Barnes, D. R. Caffrey, A. Visintin, E. Latz, B. Monks, P. M. Pitha, and D. T. Golenbock (2003). LPS-TLR4 signaling to IRF-3/7 and NF-kappaB involves the toll adapters TRAM and TRIF. *J. Exp. Med.* **198**: 1043–55.

71. Yamamoto, M., S. Sato, H. Hemmi, S. Uematsu, K. Hoshino, T. Kaisho, O. Takeuchi, K. Takeda, and S. Akira (2003). TRAM is specifically involved in the Toll-like receptor 4-mediated MyD88-independent signaling pathway. *Nat. Immunol.* **4**: 1144–50.

72. Liu, Y. J. (2005). IPC: Professional type 1 interferon-producing cells and plasmacytoid dendritic cell precursors. *Annu. Rev. Immunol.* **23**: 275–306.

73. Hemmi, H., T. Kaisho, K. Takeda, and S. Akira (2003). The roles of Toll-like receptor 9, MyD88, and DNA-dependent protein kinase catalytic subunit in the effects of two distinct CpG DNAs on dendritic cell subsets. *J. Immunol.* **170**: 3059–64.

74. Hochrein, H., B. Schlatter, M. O'Keeffe, C. Wagner, F. Schmitz, M. Schiemann, S. Bauer, M. Suter, and H. Wagner (2004). Herpes simplex virus type-1 induces IFN-alpha production via Toll-like receptor 9-dependent and -independent pathways. *Proc. Natl Acad. Sci. U S A* **101**: 11416–21.

75. Kawai, T., S. Sato, K. J. Ishii, C. Coban, H. Hemmi, M. Yamamoto, K. Terai, M. Matsuda, J. Inoue, S. Uematsu, O. Takeuchi, and S. Akira (2004). Interferon-alpha induction through Toll-like receptors involves a direct interaction of IRF7 with MyD88 and TRAF6. *Nat. Immunol.* **5**: 1061–8.

76. Honda, K., H. Yanai, T. Mizutani, H. Negishi, N. Shimada, N. Suzuki, Y. Ohba, A. Takaoka, W. C. Yeh, and T. Taniguchi (2004). Role of a transductional-transcriptional processor complex involving MyD88 and IRF-7 in Toll-like receptor signaling. *Proc. Natl Acad. Sci. U S A* **101**: 15416–21.

77. Uematsu, S., S. Sato, M. Yamamoto, T. Hirotani, H. Kato, F. Takeshita, M. Matsuda, C. Coban, K. J. Ishii, T. Kawai, O. Takeuchi, and S. Akira (2005). Interleukin-1 receptor-associated kinase-1 plays an essential role for Toll-like receptor (TLR)7- and TLR9-mediated interferon-{alpha} induction. *J. Exp. Med.* **201**: 915–23.

78. Kerkmann, M., S. Rothenfusser, V. Hornung, A. Towarowski, M. Wagner, A. Sarris, T. Giese, S. Endres, and G. Hartmann (2003). Activation with CpG-A and CpG-B oligonucleotides reveals two distinct regulatory pathways of type I IFN synthesis in human plasmacytoid dendritic cells. *J. Immunol.* **170**: 4465–74.

79. Honda, K., Y. Ohba, H. Yanai, H. Negishi, T. Mizutani, A. Takaoka, C. Taya, and T. Taniguchi (2005). Spatiotemporal regulation of MyD88-IRF-7 signalling for robust type-I interferon induction. *Nature* **434**: 1035–40.

80. Diebold, S. S., M. Montoya, H. Unger, L. Alexopoulou, P. Roy, L. E. Haswell, A. Al-Shamkhani, R. Flavell, P. Borrow, and C. Reis e Sousa (2003). Viral infection switches non-plasmacytoid dendritic cells into high interferon producers. *Nature* **424**: 324–8.

81. Smith, E. J., I. Marie, A. Prakash, A. Garcia-Sastre, and D. E. Levy (2001). IRF3 and IRF7 phosphorylation in virus-infected cells does not require double-stranded RNA-dependent protein kinase R or Ikappa B kinase but is blocked by Vaccinia virus E3L protein. *J. Biol. Chem.* **276**: 8951–7.

82. Yoneyama, M., M. Kikuchi, T. Natsukawa, N. Shinobu, T. Imaizumi, M. Miyagishi, K. Taira, S. Akira, and T. Fujita (2004). The RNA helicase RIG-I has an essential function in double-stranded RNA-induced innate antiviral responses. *Nat. Immunol.* **5**: 730–7.

83. Andrejeva, J., K. S. Childs, D. F. Young, T. S. Carlos, N. Stock, S. Goodbourn, and R. E. Randall (2004). The V proteins of paramyxoviruses bind the IFN-inducible RNA helicase, mda-5, and inhibit its activation of the IFN-beta promoter. *Proc. Natl Acad. Sci. USA* **101**: 17264–9.

84. Kato, H., S. Sato, M. Yoneyama, M. Yamamoto, S. Uematsu, K. Matsui, T. Tsujimura, K. Takada, T. Fujita, O. Takeuchi, and S. Akira (2005). Cell type-specific involvement of RIG-I in antiviral response. *Immunity* **23**: 19–28.

85. Balachandran, S., E. Thomas, and G. N. Barber (2004). A FADD-dependent innate immune mechanism in mammalian cells. *Nature* **432**: 401–5.

86. Girardin, S. E., I. G. Boneca, L. A. Carneiro, A. Antignac, M. Jehanno, J. Viala, K. Tedin, M. K. Taha, A. Labigne, U. Zahringer, A. J. Coyle, P. S. DiStefano, J. Bertin, P. J. Sansonetti, and D. J. Philpott (2003). Nod1 detects a unique muropeptide from Gram-negative bacterial peptidoglycan. *Science* **300**: 1584–7.

87. Chamaillard, M., M. Hashimoto, Y. Horie, J. Masumoto, S. Qiu, L. Saab, Y. Ogura, A. Kawasaki, K. Fukase, S. Kusumoto, M. A. Valvano, S. J. Foster, T. W. Mak, G. Nunez, and N. Inohara (2003). An essential role for NOD1 in host recognition of bacterial peptidoglycan containing diaminopimelic acid. *Nat. Immunol.* **4**: 702–7.

88. Girardin, S. E., I. G. Boneca, J. Viala, M. Chamaillard, A. Labigne, G. Thomas, D. J. Philpott, and P. J. Sansonetti (2003). Nod2 is a general sensor of peptidoglycan through muramyl dipeptide (MDP) detection. *J. Biol. Chem.* **278**: 8869–72.

89. Ogura, Y., D. K. Bonen, N. Inohara, D. L. Nicolae, F. F. Chen, R. Ramos, H. Britton, T. Moran, R. Karaliuskas, R. H. Duerr, J. P. Achkar, S. R. Brant, T. M. Bayless, B. S. Kirschner, S. B. Hanauer, G. Nunez, and J. H. Cho (2001). A frameshift mutation in NOD2 associated with susceptibility to Crohn's disease. *Nature* **411**: 603–6.

90. Kobayashi, K. S., M. Chamaillard, Y. Ogura, O. Henegariu, N. Inohara, G. Nunez, and R. A. Flavell (2005). Nod2-dependent regulation of innate and adaptive immunity in the intestinal tract. *Science* **307**: 731–4.

91. Watanabe, T., A. Kitani, P. J. Murray, and W. Strober (2004). NOD2 is a negative regulator of Toll-like receptor 2-mediated T helper type 1 responses. *Nat. Immunol.* **5**: 800–8.

92. Maeda, S., L. C. Hsu, H. Liu, L. A. Bankston, M. Iimura, M. F. Kagnoff, L. Eckmann, and M. Karin (2005). Nod2 mutation in Crohn's disease potentiates NF-kappaB activity and IL-1beta processing. *Science* **307**: 734–8.

93. Kobayashi, K., N. Inohara, L. D. Hernandez, J. E. Galan, G. Nunez, C. A. Janeway, R. Medzhitov, and R. A. Flavell (2002). RICK/Rip2/CARDIAK mediates signalling for receptors of the innate and adaptive immune systems. *Nature* **416**: 194–9.

94. Inohara, N., M. Chamaillard, C. McDonald, and G. Nunez (2005). NOD-LRR proteins: role in host-microbial interactions and inflammatory disease. *Annu. Rev. Biochem.* **74**: 355–83.

95. Bowie, A., E. Kiss-Toth, J. A. Symons, G. L. Smith, S. K. Dower, and L. A. O'Neill (2000). A46R and A52R from vaccinia virus are antagonists of host IL-1 and toll-like receptor signaling. *Proc. Natl Acad. Sci. U S A* **97**: 10162–7.

96. Stack, J., I. R. Haga, M. Schroder, N. W. Bartlett, G. Maloney, P. C. Reading, K. A. Fitzgerald, G. L. Smith, and A. G. Bowie (2005). Vaccinia virus protein A46R targets multiple Toll-like-interleukin-1 receptor adaptors and contributes to virulence. *J. Exp. Med.* **201**: 1007–18.

97. DiPerna, G., J. Stack, A. G. Bowie, A. Boyd, G. Kotwal, Z. Zhang, S. Arvikar, E. Latz, K. A. Fitzgerald, and W. L. Marshall (2004). Poxvirus protein N1L targets the I-kappaB kinase complex, inhibits signaling to NF-kappaB by the tumor necrosis factor superfamily of receptors, and inhibits NF-kappaB and IRF3 signaling by toll-like receptors. *J. Biol. Chem.* **279**: 36570–8.

98. Li, K., E. Foy, J. C. Ferreon, M. Nakamura, A. C. Ferreon, M. Ikeda, S. C. Ray, M. Gale, Jr., and S. M. Lemon (2005). Immune evasion by hepatitis C virus NS3/4A protease-mediated cleavage of the Toll-like receptor 3 adaptor protein TRIF. *Proc. Natl Acad. Sci. U S A* **102**: 2992–7.

99. Breiman, A., N. Grandvaux, R. Lin, C. Ottone, S. Akira, M. Yoneyama, T. Fujita, J. Hiscott, and E. F. Meurs (2005). Inhibition of RIG-I-dependent signaling to the interferon pathway during hepatitis C virus expression and restoration of signaling by IKKepsilon. *J. Virol.* **79**: 3969–78.

100. Fortune, S. M., A. Solache, A. Jaeger, P. J. Hill, J. T. Belisle, B. R. Bloom, E. J. Rubin, and J. D. Ernst (2004). Mycobacterium tuberculosis inhibits macrophage responses to IFN-gamma through myeloid differentiation factor 88-dependent and -independent mechanisms. *J. Immunol.* **172**: 6272–80.

101. Netea, M. G., R. Sutmuller, C. Hermann, C. A. Van der Graaf, J. W. Van der Meer, J. H. van Krieken, T. Hartung, G. Adema, and B. J. Kullberg (2004). Toll-like receptor 2 suppresses immunity against *Candida albicans* through induction of IL-10 and regulatory T cells. *J. Immunol.* **172**: 3712–18.

CHAPTER 3

MHC class I and II pathways for presentation and cross-presentation of bacterial antigens

Laurence Bougnères-Vermont and Pierre Guermonprez
Institut Curie

3.1 DENDRITIC CELLS PRIME ANTI-BACTERIAL CD4+ AND CD8+ T CELLS IN VIVO

It is now widely accepted that dendritic cells (DCs) are crucially required for the priming of T cell responses[1,2]. Major histo-compatibility complex (MHC) class I and class II presentation pathways ensure the priming of CD8+ and CD4+ T cell, respectively. They thus represent major checkpoints for the induction of adaptive protective immunity toward intracellular bacteria. In this chapter, we will focus on the basic cell biology and physiological regulation of these pathways in the context of bacterial infection. In accordance with the literature, we will refer to "cross presentation" for MHC class I pathways involved in the presentation of non cytosolic antigens.

Listeria[3,4], *Mycobacteria* [5–7] and *Salmonella*[8–12] were shown to actually infect DCs *in situ*. Some studies have addressed the capacity of DCs purified from infected animals to activate in vitro T cells specific for bacteria-encoded antigens. Intravenous infection of mice with *Salmonella*[10] and with *Mycobacterium bovis* BCG (bacillus Calmette-Guérin)[7] leads to the infection of both spleen DC subsets (CD8α− CD11c+ and CD8α+ CD11c+). Both subsets display some MHC class I and II complexes formed after the processing of bacteria-encoded antigens[7,10]. Bacterial infection may also promote apoptosis, resulting in the delivery of bacterial antigens to DCs upon the phagocytosis of infected apoptotic bodies (see Section 3.5)[13].

Whatever may be the mechanism (infection or dead cells cross presentation), the absolute requirement of DCs to induce anti-bacterial T cell priming was elegantly demonstrated in the *Listeria* model.

Using bone-marrow chimera, Lenz *et al.* first demonstrated that antigen presenting cells required for the priming of *Listeria*-specific CD8$^+$ T cells were from hematopoietic origin[14]. Jung *et al.* developed a transgenic mouse expressing selectively the diphtheria toxin receptor at the surface of DCs (CD11c-DTR)[15]. DC-depleted animals upon diphtheria toxin injection were no longer capable of mounting a CD8$^+$ T cell response against *Listeria*. This study established unambiguously, at least for *Listeria*, the crucial role of DCs in anti-bacterial T cell priming in vivo.

3.2 MHC CLASS I PATHWAYS

3.2.1 The cytosolic pathway for MHC class I presentation

Basic mechanisms of the cytosolic pathway (Figure 3.1)

MHC class I molecules bind and present peptides around ten amino acids to CD8$^+$ T lymphocytes. These peptides are generally derived from cytosolic proteins and generated upon the coordination of various cytosolic and noncytosolic protease activity. Defective ribosomal products (DRIPs) may provide the major source of antigens for the cytosolic pathway[16]. Ubiquitinylation of proteins or DRIPs target them to the proteasome for proteolytic degradation. The proteasome complex play a major role in the cytosolic proteolytic process[17]. Proteasomes are generally required for generation of the C-terminus of the peptides that are found loaded on MHC class I molecules. Under transcriptional activation (by IFNs, for example), three proteasomal subunits are replaced by their "immunological" homologues LMP2, LMP7 (encoded in the MHC) and MECL1 thus forming immunoproteasome complexes. Immunoproteasome formation may further increase the generation of "presentable" peptides by promoting the cleavage after basic or hydrophobic residues. TPPII and other cytosolic aminopeptidases participate in some cases to peptide trimming[18]. Then, peptides derived from cytosolic proteolysis are translocated into endoplasmic reticulum (ER) lumen by transporters associated with antigen processing (TAP). In the ER lumen they can undergo further N-terminal trimming by aminopeptidases such as endoplasmic reticulum aminopeptidase (ERAP)1 or ERAP2[17,18]. Optimal length peptides are then associated to MHC class I molecules by the MHC class I complex loading[19]. The MHC-encoded tapasin chaperone acts as a molecular bridge between the neosynthesized MHC class I and TAP transporters. Chaperones, like calnexin and calreticulin, and oxydoreductases like Erp57, cooperate for the assembly of the trimeric complexes containing MHC class I heavy chain, β2-microglobulin and antigenic peptides. Consequently, MHC class I-peptides traffic along the secretory pathway to cell surface. Although submitted to transcriptional regulation (notably by IFNs), this pathway is operative in most cell types allowing the detection of intracellular pathogens such as viruses and bacteria.

DCs specific features of the cytosolic pathway

Unlike many other cell types, immature DCs express significant levels of immunoproteasomes[20]. Upon LPS-triggered maturation, Pierre's laboratory characterized the transient appearance of agreggosome-like structures, the DALIS (dendritic cells aggresome-like induced structures), in DCs induced to mature by LPS[21,22]. DRIPs are targeted toward these organized and dynamic structures that seem to exclude long-lived proteins[21]. Based on this observation, the authors proposed that this process may reflect a transient shutdown of the bulk production of endogenous peptides by DRIP processing, thus favoring the loading of cross-presented peptides. The transient shutdown of DRIP processing may also favor proteasomal processing of bacterial proteins secreted in the cytosol of infected cells. In support of this, DALIS are also induced in bacteria-faced macrophages[23]. Later stages of DC maturation are characterized by the upregulation of the MHC class I presentation machinery including immunoproteasomes, TAP, β2 microglobulin and MHC class I heavy chain[24]. At the level of MHC class I trafficking, Ackerman *et al.* showed that immature DCs transiently retain MHC class I in Golgi-associated pools[25], whereas the induction of DC maturation accelerates their delivery to the cell surface. Unlike MHC class II, MHC class I half life is modestly increased during DCs maturation[25–28]. Production of MHC class I-peptide complexes is sustained in mature DCs[26].

Bacterial antigens secreted by live, cytosol-invading bacteria enter the cytosolic pathway

Intracellular bacteria such as *Listeria* and *Shigella* enter the cytosol of invading cells where they replicate. The *Listeria* listeriolysin LLO triggers *Listeria*-containing vacuole lysis which results in escape of the whole bacterium into the cytosolic compartment[29]. CD8+ T cells are major components of the adaptive immune response against *Listeria*. Pioneering cellular studies from Pamer's lab analyzed the pathways leading to the presentation of bacteria-derived peptides by MHC class I molecules[29,30]. Using a biochemical approach, they identified and quantified three

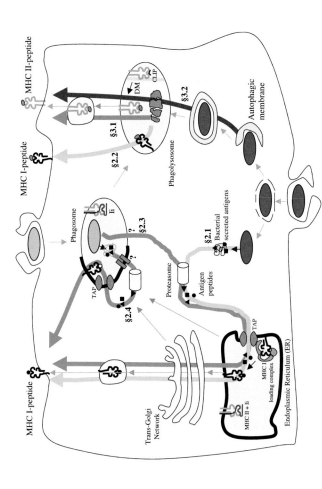

Figure 3.1. MHC I and MHC II presentation pathways for bacterial antigens. The "classical" cytosolic MHC-I pathway is involved in presentation of bacterial secreted antigens delivered by live cytosolic bacteria (Section 3.2.1) and cross-presentation of vacuolar bacteria-derived antigens (Section 3.2.3). The ER-phagosome TAP-dependent pathway for MHC-I cross-presentation may be used in the case of ER-containing bacterial vacuoles (Section 3.2.4). The TAP-independant phagolysosomal pathway is used for MHC-I-restricted cross-presentation of some vacuolar bacteria-derived antigens (Section 3.2.2). The phagolysosomal pathway is involved in MHC-II presentation of vacuolar bacteria-derived antigens (Section 3.3.1) and may be used in the case of cytosolic bacteria sequestered by the autophagic pathway (Section 3.3.2).

major epitopes presented by H-2K^d MHC class I molecules at the surface of infected cells. Two of them belong to the p60 secreted protein, the third one belongs to the LLO. As shown by pharmacological inhibition, the generation of these epitopes depends on proteasomal degradation of *Listeria* neosynthesized proteins. Villanueva *et al.* quantified the efficiency of p60 presentation: depending on the epitope, 3 or 30% of the p60 gave rise to peptide-MHC complexes[31,32]. These yields are much higher than those found for a substrate targeted to complete degradation: in this case 1 out of 2000 proteins gave rise to a peptide–MHC complex[33]. This difference could rely on methodological discrepancies between the two studies. Alternatively, the cytosolic innate response triggered by bacteria may control the efficiency of antigen processing pathways.

Moreover, genetic insertion of N-terminus residues that target the ubiquitin proteasome pathway decreased p60 protein half-life within the cytosol, and consequently increased the presentation of epitopes derived from it[34]. This elegant study demonstrated that bacterial protein proteasomal degradation by the host-cell proteasome follow the same rules as host cell proteins.

Do nonsecreted antigens enter the cytosolic pathway?

In a major study, Shen *et al.* investigated the role of *Listeria* antigen location to T cell immunity[35]. Both secreted and nonsecreted antigens were shown to suitably activate T cells in vivo suggesting that both were efficiently presented by DCs in some way. Strikingly, only CD8$^+$ T cell responses against secreted antigens were protective, suggesting that most of somatic infected cells were not capable of processing nonsecreted antigens. Several hypotheses may explain why DCs can prime CD8$^+$ T cells against nonsecreted antigens. First, DCs may have a selective ability to kill and degrade a part of cytosolic *Listeria*. This process may be linked to the autophagic pathway that has recently been identified as a defense pathway against cytosolic bacteria[36–38]. Second, a subset of DCs may limit the access of *Listeria* to the cytosolic compartment and deliver *Listeria*-derived antigens to the phagosomal cross-presentation pathway (see Sections 3.2.2 and 3.2.3)[38]. Accordingly, electronic microscopy studies performed in macrophages clearly established that not all bacteria gained access to the cytosol[39]. Third, phagocytosis of dying infected cells by DCs may also deliver *Listeria*-derived antigens to the phagosomal cross-presentation pathway[39]. A recent report from Tvinnereim *et al.* supports the latter possibility (see Section 3.5)[40].

Manipulation of bacterial pathways for antigen delivery to the cytosolic pathway

The ectopic recombinant expression of the membranolytic LLO was exploited to favor the cross presentation of antigens expressed by bacteria, such as *E. coli*[41] or *M. bovis* BCG[42], that remain within vacuoles.

Vacuolar bacteria derived virulence factors are actively injected in the cytosol to subvert host cell trafficking function. Russman *et al.* demonstrated that genetic insertion of a model antigen into the *Salmonella sptP* gene, which encodes a protein injected into host cell cytosol by a type III secretion system, can lead to productive antigen presentation by a TAP-dependent pathway[43].

In addition, bacterial toxins endowed with intrinsic translocation activities were used as a Trojan horse to deliver recombinant vaccinal T cell epitopes in the cytosolic pathway of antigen presenting cells, including DCs[44].

3.2.2 The phagosomal pathway for cross presentation

A non-cytosolic pathway for vacuolar bacteria

Several bacteria restrict their growth to intracellular vacuoles. This is a way to escape to bactericidal effector of humoral immunity such as antibodies and complement. For a long time, it was thought that the MHC class I pathway was restricted to cytosol-invading bacteria or viruses. Pioneering studies from Harding, Pfeifer and Wick led to the discovery of an "alternative pathway" for MHC class I loading in activated macrophages[45]. These authors developed recombinant *E. coli* and *Salmonella* strains expressing the chicken ovalbumin OVA model antigen. Infected macrophages presented efficiently the immunodominant OVA peptide on MHC class I molecules. Unlike the "classical" cytosolic pathway, this alternative pathway was (i) resistant to proteasome inhibition, (ii) also observed in TAP deficient macrophages, and (iii) insensitive to Golgi disruption by BFA treatment. Phagolysosomal proteolysis and loading of antigenic peptides onto a post-Golgi pool of MHC class I thus characterize this cross-presentation pathway. However, all these experiments were performed in activated macrophages raising the question of the relevance of this pathway in DCs. Recent studies confirmed that this pathway was used for the cross presentation of recombinant *M. bovis* BCG expressing the OVA antigen by both bone marrow- and spleen-derived DCs[46]. In support of this, *M. bovis* BCG was shown to act as an efficient vector for the delivery of antigens to the phagosomal MHC class I pathway[47–49].

The comparison of various types of particles and the decrease in antigen presentation in TAP-deficient cells suggest that the phagosomal pathway is of low efficiency and biased toward highly abundant phagosomal antigens[50,51]. However, we note that the availability of MHC class I for post-Golgi loading may be lower in TAP-deficient cells used to define this pathway[52–55]. Indeed, in the absence of TAP, empty MHC class I molecules accumulate in the ER. Therefore, the TAP-independent pathway may have been underestimated in DCs due to this experimental limitation.

Post-Golgi MHC class I trafficking and loading

Initially, Pfeifer *et al.* showed that infected macrophages may support the transfer of peptides to bystander antigen presenting cells[45]. This process called "regurgitation" supports the idea that degradative macrophages may cooperate in some way with DCs to generate bacteria-derived peptides. However, there is yet no evidence of the existence of such a process in vivo. Thus, the current view is that the phagosomal pathway involves MHC class I loading in bacterial phagosomes or other vesicular post-Golgi compartments.

Do MHC class I traffic through the endocytic system in DCs?

Early studies demonstrated that MHC class I molecules surface level is highly increased during DC maturation, due to a major increase in their neosynthesis rate and to a modest stabilization of their half-life[25–28]. In CD34-derived human Langerhans cells and in blood DCs, but not in monocyte-derived DCs, MHC class I together with HLA-DM and HLA-DR are found in late multivesicular MHC class II-containing endosomes[56,57]. In maturing mouse DCs, MHC class I molecules transiently colocalize with MHC class II and B7 molecules in non-lysosomal compartments, probably trafficking to cell surface[58]. These endocytic population of MHC class I may originate from the Golgi or from cell surface through endocytosis. A recent paper from Jefferies's laboratory addressed the role of endosomal MHC class I trafficking by producing transgenic mice expressing Kb molecules mutated on a conserved tyrosine in their cytosolic domain[59]. Confirming previous studies showing that this tyrosine is required for MHC class I endocytosis[60,61], the endocytic localization of tyrosine-mutated Kb molecules was reduced in immature transgenic DCs. This result suggested that the contribution of Golgi-to-endosomes trafficking to the MHC class I endosomal pool, if any, was limited. Strikingly, cross presentation of exogenous proteins or viruses was markedly decreased in these mice. Using a monoclonal antibody specific for OVA peptide-MHC class I complexes, the authors showed that the tyrosine mutation affected

MHC class I loading in endocytic positive compartments. Whether endocytosis of MHC class I controls bacterial antigen cross presentation by DCs remains to be determined.

The exchange of peptides already bound to MHC class I with phagosome-derived peptides has been proposed as a mechanism for endocytic MHC class I loading. However, phagosomes usually undergo acidification and the consequence of a low pH environment on MHC class I peptide exchange is not fully understood.

The pathway allowing MHC class I, once loaded, to reach cell surface has not yet been well characterized in DCs but it is tempting to speculate that it may share, at least in part, some transport pathways used by MHC class II[58] such as the tubulo-vesicular structures induced upon DC maturation[62–64].

3.2.3 Phagosomal antigens can access to the cytosolic pathway

As stated before, most of the pioneering studies using recombinant bacteria expressing the OVA antigen were done in activated macrophages with bacteria, such as *E. coli* or *Salmonella*, that do not actively disrupt their vacuole. The development of DC culture protocols led to the reconsideration of these issues in DCs using the same recombinant bacteria. Strikingly, the cellular pathways used for cross presentation were proved to be different in DCs compared to macrophages. Rescigno[27] and Svensson[65] identified a "classical" cytosolic pathway for the cross presentation of vacuolar OVA-expressing bacteria (*Streptococcus gordonii, Salmonella*). Indeed the activation of OVA-specific CD8$^+$ T cells was resistant to proteasome inhibition and TAP-dependent. Accordingly, *Salmonella* infected DCs did not exhibit the ability to "regurgitate" peptides to neighboring cells. These results led to the conclusion that antigens from vacuolar bacteria gained access to the cytosol in some way where they could meet proteasomal degradation.

How exogenous antigens transit from bacterial vacuoles to the cytosol?

One may wonder whether a proportion of vacuolar bacteria may not actually gain access to the cytosol and deliver antigens to the cytosolic pathway. This hypothesis may be supported by reports of *Salmonella*[66] or *Mycobacterium*[67,68] cytosol invasion. Moreover, the recent identification of a repair system for cellular membranes injured by bacteria secretion systems suggests that the maintenance of the vacuolar membrane integrity is a dynamic and regulated phenomenon[69]. Supporting this view, the SifA *Salmonella* protein was shown to be involved in the maintenance of vacuole integrity[70]. Thus, even in the absence of global disruption, transient permeability may be activated[71].

Numerous studies support the existence of a cell transport process linking phagosomal lumen to the cytosol. Pioneering studies from Rock's laboratory demonstrated in macrophages that phagocytosis of micron-sized beads triggered specifically the egress of proteins from the phagosome to the cytosol. Indeed, when beads are coated with the OVA antigen this leads to its delivery to the cytosolic pathway, and when they are coated with the cell impermeable gelonin toxin this leads to protein synthesis inhibition[72]. Other groups have suggested that phagosome-to-cytosol pathway was triggered by phagosome rupture due to the solid nature of latex beads[73,74]. This was confirmed by the studies from Norbury *et al.* showing that growth factor-induced macropinocytosis in macrophage[75] and constitutive macropinocytosis in immature DCs[76] triggered the access of soluble proteins to the cytosolic pathway. Later, it was shown that, in DCs, transport from the endocytic pathway to the cytosol is selective for the size of the molecules transported[77]. By contrast to antigens, endosomal resident proteins such as active cathepsin D[77] or L[28] remain confined to the endocytic pathway. This selectivity may rely on a specific sorting of exogenous protein antigens or, more conceivably, to the non-specific release into the cytosol of compartments that are not enriched with active cathepsin D or L. The molecular machinery involved is still unknown but may require communication with the ER (see Section 3.2.4).

Of note, antigen transport from phagosomes to cytosol is more efficient in DCs than in macrophages. This could explain why the same bacterial antigen are presented by a TAP-dependent mechanism in DC and a TAP-independent mechanism in macrophages[77].

The TAP dependent MHC class I pathway for exogenous antigens is highly regulated by a Toll-like receptor (TLR). First, TLR signalling promotes rapidly the macropinocytic uptake of soluble antigens through the rapid remobilization of intracellular actin pools from podosome to forming ruffles[78]. Accordingly, Gil-Torregrossa *et al.* showed that DC pre-incubation with LPS efficiently promotes the uptake of immune complexes and their introduction into the cytosolic pathway[24]. Delamarre *et al.* also reported that LPS promoted cross presentation of soluble antigen in early-activated DCs[28]. The impact of TLR2–4 signalling on the dynamics of phagolysosomal maturation has been studied in macrophages but it remains a controversial issue[79,80]. How TLR signalling may influence cross presentation of bacteria encoded antigens in DCs remains unknown.

3.2.4 ER-phagosome fusion and cross presentation

ER-phagosome fusion defines a TAP-dependent "cross-presentation" compartment

As described in the previous section, most of the cross-presentation pathways described in infected DCs were proteasome- and TAP-dependent thus suggesting that ER was actually the compartment where MHC class I loading takes place. However, morphological evidence directly supporting this interpretation was slight. The characterization of ER recruitment to phagosomal membranes challenged this view and led to the proposal of a new model for cross presentation.

A pioneering paper from Desjardin's lab established that phagocytosis of inert as well as bacterial particulate substrate actually involved ER-phagosome fusion. Gagnon *et al.* performed morphologic analysis of latex beads containing phagosomes in murine macrophages using electronic microscopy[81]. Phagosomal membranes contain numerous domains positive for glucose-6-phosphatase or calnexin, two ER markers. Moreover, proteomic analysis of fractionated phagosomes revealed that they contain numerous ER proteins[82]. Morphological analysis of phagocytic cups suggested that ER recruitment may occur by direct fusion between the ER and the plasma membrane[81]. The same observations were obtained in DCs[83,84].

What is the impact of ER recruitment to the proteasome- and TAP-dependent pathway?

First, ER-derived membranes may provide a machinery for the export of phagosomal antigens to the cytosolic proteasome-dependent pathway. Indeed, numerous misfolded proteins of the ER lumen are exported to the cytosol, where they are deglycosylated, ubiquitinylated and finally targeted to proteasome degradation, a process called ER-associated degradation (ERAD) (for a review see Ref. 85). Importantly, ERAD can promote the presentation of lumenal proteins from ER by a proteasome and TAP-dependent pathway. ERAD has been coopted by viruses to trigger MHC class I breakdown[86]. In this case, co-precipitation experiments suggest that Sec61 may act as transporter for delivering ERAD substrates to the cytosol[87] and possibly some bacterial toxins whose translocation to the cytosol starts in the ER[85]. However, other studies indicate that the US2/11-mediated retrotranslocation may bypass protein unfolding, questioning the contribution of Sec61 whose substrates may be limited to fully unfolded proteins[88]. Recent results from the groups of H. Ploegh[89] and T. Rappoport[90], implicate the membrane protein Der1p in the degradation of MHC class I heavy chain triggered by US11. Der1p associates with both US11 and with VIMP, a novel membrane protein that recruits the cytosolic p97 AAA ATPase required for MHC class I retro-translocation and proteasome targeting of ER-derived substrates[90,91], Phagosome-associated proteasomes may be involved in the degradation of phagosomal antigens targeted to the cytosol[92], The contribution of these pathways to cross presentation remains to be established.

Second, the ER-phagosome fusion model led to the hypothesis that phagosomes may represent competent organelles for TAP-dependent loading of proteasome derived peptides. This hypothesis received strong experimental support. Two studies showed that purified phagosomes indeed support ATP-dependent TAP-mediated peptide import[83,84]. Moreover, phagosomal MHC class I is integrated in a loading complex encompassing chaperones and oxydoreductases that participate in the assembly of peptide-MHC complexes (calreticulin, Erp57 and tapasin), which are physically linked to TAP[19,84]. The glycosylation status of the phagosomal MHC class I associated to TAP confirms that it is mostly from pre-Golgi origin[84]. Moreover it could be hypothesized that MHC class I trafficking from cell surface through its internalization motif may also associate with loading complexes from ER origin[59]. Whatever its origin, in vitro experiments demonstrated that phagosomal MHC class I is accessible to cytosolic peptides imported by phagosomal TAP transporters[83]. MHC class I molecules loaded with a peptide derived from phagocytosed antigen were detected in phagosomes, using both antigen-specific T cells and a monoclonal antibody specific for the same complex[83]. As for cell surface appearance of MHC class I-peptide complexes, phagosomal peptide-MHC class I complexes were not detected when the cells were treated with proteasome inhibitors, indicating that peptide generation occurred in the cytosol. The contribution of this loading pathway, as compared to ER loading, to cross presentation remains unclear. MHC class I-peptide complexes were found to be enriched in antigen-bearing phagosomes, as compared to control phagosomes in the same cells[83]. These results suggest that MHC class I loading is favored in antigen-bearing phagosomes, as compared to other compartments such as ER or phagosomes that do not contain antigen.

How do peptide-MHC class I complexes loaded in phago-ER-some reach cell surface?

MHC class I may travel through the Golgi or directly to cell surface via an endocytic recycling route. The partial inhibition observed with the

Golgi-disrupting agent BFA suggest that the second alternative is possible[92]. In support of this, Ackerman *et al.* observed that a portion of cell surface MHC class I retains a pre-Golgi glycosylation status, suggesting that it bypassed Golgi trafficking[25,84].

The ER-phagosome model give an explanation for the high efficiency of phagocytic cross presentation. However, it does not explain how soluble antigens acquired by constitutive macropinocytosis could access TAP-dependent MHC class I presentation[76]. Elegant studies from Cresswell's laboratory have recently answered this question. Ackerman *et al.* showed that the cytomegalovirus soluble protein US6, known to inhibit TAP function by interacting with its luminal domain, inhibited the cross presentation of soluble OVA[84]. These data demonstrated that soluble antigens have access in some way to compartment presenting ER components. Strikingly, US6 also inhibited significantly the TAP-dependent presentation of endogenous proteins[93]. Moreover, exogenous β2-microglobulin localized rapidly to the perinuclear ER, associated with MHC class I heavy chains, and rescued both cell surface class I expression in β2 microglobulin $^{-/-}$ cells and presentation of endogenous antigens[93]. These data give evidence that soluble antigens can access most of ER lumen in DCs where they gain access to the cytosolic pathway.

ER recruitment to bacterial vacuoles: a role in cross presentation?

Legionella pneumophila and *Brucella abortus* are able to create replicative organelles within macrophages, by actively inducing ER recruitment to the bacterial phagosome. The control of ER-phagosome biogenesis depends in both cases on a functional type IV secretion system that injects virulence factors in the host cell cytosol (the Dot/Icm system for *Legionella*[94] and the VirB system for *Brucella*)[95,96]. Even though their ER-derived replicative vacuoles present morphological similarities, *Legionella* and *Brucella* trigger active ER-phagosome fusion through distinct pathways.

Legionella avoid fusion with early endosomes and acquires ER membranes through the subversion of ER to Golgi vesicular trafficking. As shown by Roy's lab, Arf1 is activated by the Dot/Icm-injected RalF guanine nucleotide exchange factor that participates in its recruitment to *Legionella* phagosomes[97]. There, Arf1 plays a crucial role in sustained bacterial growth. The small GTPase Rab1 and the SNARE Sec22, are required for the subversion of the formation of *Legionella* replicating vacuole[98]. Interestingly, yeast Sec22 was shown to support the fusion of liposomes with plasmalemma-derived SNAREs, thus suggesting that it may regulate ER-phagosome fusion[99]. This idea recently received experimental evidence in mammalian cells: the phagocytose of 3 μm latex beads by J774 macrophages is blocked by microinjection of anti-Sec22b antibodies or expression of dominant negative forms of Sec22b[100].

After the acquisition of certain lysosomal markers[95], *Brucella* containing-vacuoles intercept the host secretory pathway at a slightly earlier point as compared to *Legionella* by a process that requires the activity of the Sar-1 small GTPase[95]. The relevance of ER-derived membrane acquisition for cross presentation of *Brucella* or *Legionella* antigens remains to be tested.

3.3 MHC CLASS II PATHWAYS

3.3.1 The endo-lysosomal pathway for MHC class II presentation

The formation of peptide-MHC class II complexes for presentation to CD4$^+$ T cells is an endocytic process associated to endosomal proteolysis both at the level of antigen processing and of MHC class II maturation pathway. Here, we will only summarize briefly tremendous work that has been nicely reviewed elsewhere[101,102].

Neosynthesized MHC class II α and β chains assemble in the endoplasmic reticulum (ER) with a non-polymorphic transmembrane protein called invariant chain (Ii). Ii triggers the formation of nonameric complexes encompassing three MHC class II dimmers and three Ii molecules. A cytoplasmic domain of Ii addresses these complexes from the ER to the endocytic pathway through the Trans-Golgi Network. A minor portion of neosynthesized MHC class II gain direct access to the cell surface before being internalized into endosomes. Successful antigen presentation relies on breakdown and removal of Ii, which must be proteolyzed upon arrival to endosomes to free the class II-peptide binding groove for exchange with the antigenic peptide. Peptide exchange is catalyzed by the chaperone H2-DM whose activity is triggered at acidic pH. In addition, the last step of Ii proteolysis liberates MHC class II from the cytosolic tail of Ii, which contains the endosomal retention motif and thereby allows MHC class II-peptide complexes to reach the cell surface. This step of Ii degradation is under the control of the cysteine protease, cathepsin S, in both DCs and B lymphocytes. Antigenic peptides arise from the combined action of various endocytic proteases. The degradation task of these enzymes can eventually be facilitated by the acid-activated oxido-reductase, GILT, which contributes to antigen unfolding by breaking di-sulfide bonds.

Most of these hydrolases display optimal activity under the acidic conditions that are characteristic of late-endosomal/lysosomal compartments.

In DCs, the MHC class II pathway is submitted to a tight developmental regulation. In association with their tissue localization, immature DCs exhibit a high ability to capture antigens by various endocytic processes such as phagocytosis, macropinocytosis or receptor-mediated endocytosis. Immature DCs may form special endocytic retention compartments devoid of V-ATPase thus retaining internalized antigens intact for further processing in mature DCs[103]. Protease inhibitors may also dampen proteolytic activity in immature DCs[104]. Immature DCs display a reduced lysosomal acidification, associated with a limited antigen proteolysis and Ii cleavage. Cell surface MHC class II are also efficiently endocytozed[102]. As a result, MHC class II are mostly intracellular and the ability of immature DCs to process antigens is low, especially for the most stable proteins.

Upon induction of maturation by LPS, DCs undergo a coordinated set of modifications associated with their migration toward lymphoid organs. One major event is the acidification of late endocytic compartment and activation of lysosomal function, promoting both antigen and Ii processing[105]. This event is triggered by the recruitment and docking of the V-ATPase-cytosolic subunit V1 to the lysosomal membrane-associated V0 subunit, but the molecular mechanisms regulating this event remain to be established[105]. At the level of trafficking, MHC class II are targeted toward tubular compartments linking the late endocytic pathway to cell surface[62]. In mature DCs, the endocytosis of cell surface MHC class II is deeply inhibited and as a result, high levels of MHC class II accumulate at the cell surface[106]. Moreover, in mature DCs, the half life of MHC class II is greatly increased thus sustaining prolonged antigen presentation to CD4[+] T cells in the lymph nodes. Thanks to all these coordinated modifications, once matured, DCs display the striking ability to present antigens, even several days after endocytic uptake, a phenomenon termed as "antigenic memory" (reviewed in Ref. 101). In conclusion, developmental regulation of DC function allows the precise coordination of antigen transport and degradation in the course of DC migration from peripheral organs where they capture antigens, to lymphoid organs where they present antigens to T cells[101,107].

Important progress in the characterization of phagosome function in MHC class II presentation was performed in macrophages. Ramachandra et al. showed that phagosomes carrying inert beads matured toward phagolysosomes and acquired MHC class II, Ii and DM. Nascent

MHC class II trafficking from Golgi and recycling MHC class II derived from cell surface were found in phagosomes. These phagosomes are degradative organelles where antigen and Ii are efficiently processed[108,109]. Thus, they have all the features of a MHC class II loading compartment. However, this does not exclude that loading may take place in other late endocytic compartments upon antigen or peptide vesicular transport. To resolve this issue, Ramachandra *et al.* developed a unique experimental approach to identify specific MHC class II-peptide in the membrane of fractionated organelles. Using this technique, the authors demonstrated that phagosomes were competent for MHC class II loading of peptides derived from phagocytozed antigens. MHC class II loading was not detectable in non-phagosomal organelles, showing that antigen (or peptide) export to other organelles was not a significant phenomenon, at least for MHC class II loading. Phagosomal MHC class II loading pathway use primarily nascent MHC class II derived from the Golgi[108,109]. The same authors applied this approach in order to analyze the MHC class II-restricted presentation of a peptide derived from the Ag85B secreted by *Mycobacterium tuberculosis (Mtb)*[110]. This study formally established that the *Mtb*-containing phagosome is actually the major organelle for specific MHC class II loading in activated macrophages even if some escape mechanisms are set up by live bacteria to inhibit phagolysosomal maturation and the MHC class II pathway (see Section 3.4)[110].

3.3.2 Cytosol derived antigens can access the MHC class II pathway

A strong CD4 T cell response has been reported following mice infection with the intracellular bacteria *Listeria monocytogenes*. Moreover, macrophages and DCs isolated from these mice efficiently present *Listeria*-derived epitopes on MHC class II[111,112]. Therefore, despite the intracytosolic location of *Listeria*, its secreted proteins have a direct access to the endocytic MHC class II presentation pathway. A possibility would be that DCs acquire bacterial-derived antigen by phagocytosis of dead bacteria or of *Listeria*-infected cells. In vitro experiments by Skoberne *et al.* validated this hypothesis, but also showed that the direct infection of DCs and macrophages can lead to the presentation of *Listeria*-derived epitopes on MHC-II molecules[113].

The fact that bacterial cytosolic antigens can be presented on MHC class II is not so surprising. Indeed, a large proportion of peptides bound to MHC class II, derived from endogenous proteins residing in the cytosol (reviewed in Ref. 101). However, the underlying mechanisms allowing cytosolic antigens access to a compartment competent for loading on MHC class II are poorly characterized at the cell biology level.

Despite exceptions[114], it seems that this pathway is independent of the proteasome and of the ER resident peptide transporter TAP. Since the process of peptide loading on MHC class II (Ii release, DM activation) require endosomal acidification, the access to endocytic pathway is more likely to be required. Indeed, the direct transport of peptides derived from cytosolic proteins into late endosomal compartment positives for LAMP-1 and cathepsin-D was reported[115].

But how is the connection between the exocytic and the endocytic pathways achieved?

A system called chaperone-mediated autophagy, activated upon metabolic stress, triggers the transfer across membranes of cytosolic proteins containing a particular targeting sequence, directly into endosomes and lysosomes[116]. Interestingly, the two major constituents of this system, hsc70 and Lamp2a, were recently reported to facilitate MHC class II presentation of cytosolic antigens[117]. However, there is no evidence that cytosolic bacterial antigens can be transported into lysosomes by this process.

Another cellular pathway called macroautophagy is involved in cytosolic protein turn-over and is more likely to be involved in MHC class II loading of cytosolic antigens. Indeed, this process, induced in response to cell starvation, triggers the sequestration of cytosolic components into double-membrane bound vacuoles called autophagosomes for later degradation in lysosomal compartments.

Recent studies point out a link between autophagy and the delivery of cytosolic peptides for MHC-II-restricted presentation. Indeed, Paludan et al. have shown that silencing of *Apg12*, an essential autophagy gene, by RNA interference, inhibits MHC class II-restricted presentation of the Epstein–Barr virus antigen EBNA1[118]. Moreover, Dengiel et al. have reported that induction of autophagy strongly enhances MHC-II presentation of peptides derived from cytosolic proteins[119].

Insights in the role of the autophagic pathway were made possible thanks to the identification of specific markers of mammalian autophagy (Apg5, Apg12 and LC3)[120]. Interestingly, cell biology studies exploiting these new tools revealed a link between the autophagic pathway and intracytosolic pathogenic bacteria behavior in infected cells. Indeed, autophagy appears to be a bactericidal pathway controlling the targeting of cytosolic bacteria into lysosomal compartments. As an example, the pathogenic *group A Streptococcus* is destroyed by the autophagic machinery in human cells[36]. In contrast, *Shigella flexneri* and *Listeria monocytogenes* prevent this process through the action of virulence effectors[37,38]. Therefore, the autophagic process might be a way to bring cytosolic bacterial antigens to MHC class II for loading into endocytic compartments, resulting in MHC class II presentation of bacterial epitopes at the cell surface that elicit CD4 T cell responses against pathogens.

3.4 THE INTERPLAY BETWEEN BACTERIAL VIRULENCE AND MHC CLASS I AND CLASS II PATHWAYS

DCs are permissive to infection by low virulent bacterias

MacPherson *et al.* established that gut-derived DCs, but no other cell types, carry live commensal bacteria to mesenteric draining lymph nodes after their uptake in the gut[121]. Once in lymph nodes, DC can interact directly with B cells to promote their differentiation in IgA secreting cells[121]. However, even if IgA production to commensals seems to bypass antigen presentation to T cells[122], one may speculate that the transport of live bacteria to draining lymph nodes could promote antigen presentation by MHC class I or II to T cells. Nevertheless, the elegant studies from MacPherson *et al.* strongly suggest that DCs are endowed with specific mechanisms allowing them to maintain phagocytozed bacteria alive even after the completion of the migration phase from tissues to draining lymph nodes[121].

The low bactericidal activity of DCs toward commensals is coherent with their ability to support infection by bacterial mutants devoid of essential virulence factors and unable to infect macrophages. Indeed, Niedergang and Garcia del Portillo have shown that DCs, unlike macrophages, were efficiently infected by PhoP-deficient *Salmonella* strains[9,123]. The PhoP/PhoQ system controls the coordinated expression of virulence genes involved in bacterial invasion and intracellular survival. Moreover, DCs presented *Salmonella*-derived antigens on MHC class II independently of the PhoP-regulated virulence factors[9]. The high macropinocytic activity of immature DCs promotes *Salmonella* entry independently of the *Salmonella* pathogenicity island (SPI)-1 type III secretion system absent in PhoP mutant strain. This suggests that the *Salmonella*-containing vacuoles differ in DCs and macrophages. Indeed, the lack of lysosomal glycoproteins acquisition by *Salmonella* vacuoles in DCs supports this view[123]. This underscores how DC and macrophage endocytic pathways may face bacteria to totally different constraints, probably in relation to the reduced lysosomal function of DCs (at least in the immature stage).

Bacterial inhibition of phagolysosomal function and antigen presentation in macrophages and DCs

Uptake of inert particles such as latex beads leads to the formation of a phagosome inside the host cell. Upon multiple fusion and fission events with other vesicular organelles, this phagosome progressively becomes a phagolysosome, a process called phagosomal maturation[124]. Pathogenic vacuolar bacteria invade cells through a phagocytic process and actively inhibit phagolysosome formation in order to create a niche for their replication.

For example, *Mycobacterium tuberculosis (Mtb)* actively inhibits the delivery of V-ATPase to its containing-vacuole[125] and prevents phagolysosomal maturation[126]. Deretic's laboratory had a central contribution to understand this process in which phosphoinositol-3-phosphate (PI3P) metabolism play a central role. Indeed, PI3P formation is essential for phagolysosome biogenesis as it recruits EEA1 which cooperates with syntaxin6 in the delivery of lysosomal hydrolases and V_0-ATPase from the Trans-Golgi Network to phagosomes[127]. *Mtb* cell wall glycolipid lipoarabinomannan (LAM) inhibits the activation of the PI3-kinase hVPS34 by blocking a Ca^{2+}-activated calmodulin/CaMKII signalling cascade[128,129]. Moreover live *Mtb* secretes a phosphatase called SapM that inhibit phagolysosomal maturation by hydrolyzing PI3P[130]. As PI3P also recruits the p40 subunit of NADPH-oxidase which positively regulates superoxide production, the inhibition of PI3P formation at the membrane of live *Mtb*-containing phagosomes, could also have an effect in diminishing the oxidative burst[131]. The inhibition of phagolysosomal function by live *Mtb* is associated with an inhibition of MHC class II presentation that was mostly studied in monocytes/macrophages (reviewed in Ref. 132). Ramachandra *et al.*[133] demonstrated that inhibition of MHC class II presentation involve factors resistant to certain procedure of bacterial killing except heat-killing (possibly LAM, or TLR2 agonists, see below). The inhibition by *Mtb* of PI3P-dependent trafficking from Trans-Golgi Network to *Mtb* vacuole, may also concern the recruitment of neosynthesized MHC class II and DM molecules toward *Mtb* vacuole[134]. Accordingly, Ii processing and peptide loading on MHC class II where inhibited as demonstrated biochemically by the reduction of SDS-stable peptide-loaded MHC-IIα/β dimmers in *Mtb*-infected cells[134]. This study points out the link between subvertion of cell transport pathways by *Mtb* and the inhibition of MHC class II presentation. However, the picture may be more complex as other inhibitory mechanisms seem to directly target the IFNγ-induced CIITA transactivator of the genes coding for MHC class II, Ii and DM. The activation of TLR2 by the *Mtb* 19 kD lipoprotein controls in macrophages the inhibition of transcriptional activity by mechanisms that remains to be determined (reviewed in Ref. 132). Of note, MHC class I surface levels seem to be less affected upon *Mtb* infection, perhaps in relation with the protective role of CD8$^+$ T cells in chronic infection[135].

A likewise major consequence of the inhibition of antigen presentation in infected macrophages is that the delivery of bacteriostatic cytokines provided by CD4$^+$ T cells is inhibited. However, the relevance of the phagolysosomal inhibition for the priming of CD4$^+$ (and CD8$^+$) T cells by *Mtb*-infected DCs remains to be carefully analyzed.

DC infection by *Salmonella typhimurium* was reported to inhibit the MHC class II presentation of bystander antigens through a mechanism dependent of the SPI-2 type III secretion system. Accordingly, SPI-2 coding genes expression is actually induced in *Salmonella* infected DCs but the SPI-2 system is dispensable for DC infection[136,137]. Interestingly, *Salmonella*-mediated inhibition of MHC class II pathways can be overcome by FcγR-mediated internalization of opsonized bacteria as a consequence of bacteria targeting to phagolysosomal degradation[138]. These reports establish that MHC presentation pathways in DCs can be a target for bacterial escape.

Constrained survival of virulent bacteria in DCs and the presentation of bacterial antigens

Several studies established that infected DCs can restrict the growth of intracellular bacteria without killing them. Using a segregative plasmid, Jantsch *et al.* elegantly demonstrated that *Salmonella* infecting DCs represented a live, non-dividing population[136]. Jiao *et al.* demonstrated that early after *M. bovis* BCG infection, splenic DCs efficiently present *M. bovis* BCG secreted antigens[7]. Interestingly, bacilli survive and remain stable in DCs up to two weeks after infection. These studies suggest that antigen presentation – at least for secreted antigens – may take place in chronically infected DCs carrying live bacilli. Moreover, infection of mice with a recombinant *M. bovis* BCG strain expressing the OVA antigen supports a long-lasting antigen presentation in vivo[48,49].

In vitro studies from Neyrolles's group promote the idea that human DCs do not behave as macrophages with respect to mycobacterial infection: Tailleux *et al.* showed that DCs restrict the growth of *Mtb*. In DCs, like in macrophages, live *Mtb*-containing phagosomes do not mature to phagolysosomes[139]. However, unlike in macrophages, the *Mtb* phagosome is disconnected from the early recycling pathway and this may restrict the accessibility of bacilli to exogenous nutrients such as iron or cholesterol[139]. Whether antigens derived from this vacuole may enter some presentation pathway remains to be determined. As proposed by Tailleux *et al.*, the distinct intracellular route of *Mtb* phagosome in DCs and in macrophages could result from the engagement of different receptors during phagocytosis: *Mtb* uptake by human DCs, but not by macrophages, is mediated by an interaction between *Mtb* LAM and the DC lectin-surface receptor DC-SIGN[6].

Legionella pneumophila inhibits the phagolysosomal maturation of its vacuole and actively promotes the formation of an ER-containing replicating vacuole. Studies from the Roy lab established that DCs but not macrophages restrict the growth of *Legionella*[140]. *Legionella* is able to prevent fusion of phagosome with lysosomes in both cell types through the action of the Dot/Icm system. Microscopy studies performed by Neild *et al.* in *Legionella*-infected DCs, demonstrated that phagosomes acquired ER membranes in a manner similar to what have been shown in macrophages. Moreover, despite a restricted growth, *Legionella* remains metabolically active and retain their ability to neosynthesize proteins as shown by the induction of green fluorescent protein (GFP) under an inducible promoter[140]. Strikingly, the constriction of bacterial growth does not exclude neosynthesized bacterial antigens from the class II presentation pathway.

Characterization of the transport pathways allowing the presentation of the neosynthesized bacterial antigens is of major interest. These pathways may be specially relevant for the induction of T cell immunity against secreted bacterial antigens by infected DCs carrying live bacteria during their migration to the T cell zones of lymphoid organs.

However, T cell responses against non-secreted bacterial antigens may rely on the intracellular killing and degradation of bacteria. Therefore, it would be interesting to determine whether the developmental regulation of phagolysosomal function that has been characterized for model antigens[101] is also relevant for bactericidal functions. In this case, one may wonder whether DCs may kill and process bacteria once they reached the T cell zones of lymphoid organs.

3.5 BACTERIA-INDUCED APOPTOSIS AND CROSS PRESENTATION OF BACTERIAL ANTIGENS

Bevan *et al.* discovered that cell-associated antigens of engrafted cells could be presented on the MHC of host antigen presenting cells[141].

This process was called cross priming. Numerous in vitro and in vivo studies gave some cellular basis to this process: antigen presenting cells can acquire and process cell-associated antigens from apoptotic bodies they phago-cytose[142]. As reviewed elsewhere, this process may be a major relevance for the induction of antiviral T cell responses[143]. Bacterial infection can trigger apoptosis in many cell types. This led to the suggestion that the phagocytosis of dead cells harboring bacterial antigens may participate to the induction of anti-bacterial adaptive immune responses through cross presentation. This idea received experimental first evidence in vitro in a study from Wick's laboratory[13]. Apoptotic bodies of *Salmonella* infected macrophages were efficiently phagocytozed by DCs triggering the cross presentation of *Salmonella* encoding antigens by MHC class I of the phagocytic DCs. In this model, cross presentation of bacteria infected apoptotic bodies seems to follow a "classical", cytosolic, pathway involving antigen access to the cytosol, proteasomal degradation and TAP-dependent loading[13]. This is in line with what has been characterized for cell-associated viral antigens[144–146]. The contribution of ER-mediated phagocytosis to the phagocytic uptake of dead cells and the transport pathways linking phagosome lumen to cytosol remain key issues that have not yet been documented. Several groups implicated long-lived proteins from dead cells as the antigenic substrate transported from dead cell bodies to cross presenting DCs[147–149]. Other groups sustain that HSP-mediated transfer of antigenic peptides from the dead cell to the cross presenting DCs[150–152]. Interestingly, bacteria-derived HSPs may also shuttle bacterial antigenic peptides to cross-presenting DCs[153].

Schaible *et al.* demonstrated that *Mtb*-triggered apoptosis produced some blebs harboring proteic and lipidic antigens that were efficiently uptaken and cross presented by bystander DCs on classical as well as CD1b MHC molecules[154]. However, the existence of this pathway does not exclude direct presentation by chronically infected DCs.

In vivo physiological relevance of cross presentation after uptake of dead cell carrying bacterial antigens remains ill defined.

In the case of *Salmonella*, PhoP-deficient strains do not induce either macrophage apoptosis or cross presentation in vitro[13]. In vivo, Wijburg *et al.* demonstrated that SPI-1 (controlled by PhoP) deficient strains still induce CD8$^+$ T cell activation for bacterial antigens[155]. This suggests that direct infection of DCs may be the main route for CD8$^+$ T cell priming against *Salmonella* antigens. However, other SPI-1-independent apoptotic pathways may provide dead cells harboring *Salmonella* antigens to cross presenting, phagocytic DCs.

In the case of the *Listeria* infection, indirect arguments suggest that the dead cell pathway may be of major relevance. Indeed, virulence factors trigger the rapid apoptosis of infected DCs both in vitro and in vivo[29]. Using adoptive transfer experiments to monitor antigen presentation in vivo, Wong *et al.* elegantly demonstrated that CD8[+] T cells are stimulated during the first days of infection[156]. However, the quick disappearance of antigen presenting cells is not only due to *Listeria*-induced apoptosis since activated CD8[+] T cells also participate in the process[156]. Therefore, these studies do not rule out either direct infection or cross presentation as main priming mechanisms. Geginat's laboratory successfully modelized in vitro the cross presentation of *Listeria* derived epitopes by both MHC class I and II[113,157]. Interestingly, the process may be of special relevance for MHC class II presentation of *Listeria*-derived epitopes that could not gain efficiently access to phagolysosomes otherwise (see Section 3.3.2). Interestingly, neutrophils may play a role in this process by providing numerous infected dead cells carrying listerial antigens to cross-presenting DCs. In support of this hypothesis, Tvinnereim *et al.* have shown that dead neutrophils serve as efficient substrates for cross presentation both in vitro and in vivo[40]. The relevance of this pathway is demonstrated by in vivo experiments in which neutrophil depletion deeply inhibits the priming of CD8[+] T cells specific for non-secreted antigens[40]. Strikingly, the priming of T cells specific for secreted antigens is not affected in these conditions. Although indirect, these data argue for a major role of cross presentation at least in the case of non-secreted listerial antigens.

3.6 CONCLUSION: BACTERIAL COMPARTMENTALIZATION AND MHC PRESENTATION PATHWAYS

Historically, the MHC-II pathways were primarily thought to be more or less restricted to endocytically acquired antigens whereas the MHC class I was believed to be assigned to the presentation of cytosol-derived antigens. Cell biology studies clearly demonstrated that this dichotomy is no more relevant in face of the complexity and variety of cellular transport pathways. At the theoretical level, it can be postulated that every bacteria, independently of its intracellular location, may provide antigens presented by MHC class I and class II pathways. This is of outstanding interest given the cooperation of CD8[+] and CD4[+] T cells in mounting efficient protective anti-bacterial memory T cell responses. However, the precise interplay between bacterial virulence factors, host anti-bacterial pathways and

MHC antigen presentation pathways remains to be better defined in physiologically-relevant DC subsets.

ACKNOWLEDGEMENTS

L. B.-V. was supported by the Association pour la Recherche contre le Cancer, and P. G. was supported by the CNRS. The authors thank Ana-Maria Lennon-Duménil and Stéphanie Hugues for critical reading of the manuscript. Due to space limitation we could only cite a fraction of the published work, which does not undermine the great value of uncited studies.

REFERENCES

1. Banchereau, J. *et al.* (1998). *Nature* **392**(6673), 245–52.
2. Guermonprez, P. *et al.* (2002). *Ann. Rev. Immunol.* **20**, 620–67.
3. Kolb-Maurer, A. *et al.* (2000). *Infect Immun.* **68**(6), 3680–8.
4. Pron, B. *et al.* (2001). *Cell Microbiol.* **3**(5), 331–40.
5. Bodnar, K. A. *et al.* (2001). *Infect. Immun.* **69**(2), 800–9.
6. Tailleux, L. *et al.* (2003). *J. Exp. Med.* **197**(1), 121–7.
7. Jiao, X. *et al.* (2002). *J. Immunol.* **168**(3), 1294–301.
8. Hopkins, S. A. *et al.* (2000). *Cell Microbiol.* **2**(1), 59–68.
9. Niedergang, F. *et al.* (2000). *Proc. Natl Acad. Sci. U S A* **97**(26), 14650–5.
10. Yrlid, U. *et al.* (2002). *J. Immunol.* **169**(1), 108–16.
11. Svensson, M. *et al.* (2000). *Infect. Immun.* **68**(11), 6311–20.
12. Johansson, C. *et al.* (2004). *J. Immunol.* **172**(4), 2496–503.
13. Yrlid, B. U. *et al.* (2000). *J. Exp. Med.* **191**(4), 613–24.
14. Lenz, L. L. *et al.* (2000). *J. Exp. Med.* **192**(8), 1135–42.
15. Jung, S. *et al.* (2002). *Immunity* **17**(2), 211–20.
16. Yewdell, J. W. *et al.* (2003). *Nat. Rev. Immunol.* **3**(12), 952–61.
17. Rock, K. L. *et al.* (2004). *Nat. Immunol.* **5**(7), 670–7.
18. Kloetzel, P. M. (2004). *Nat. Immunol.* **5**(7), 661–9.
19. Cresswell, P. (2000). *Traffic* **1**(4), 301–5.
20. Macagno, A. *et al.* (1999). *Eur. J. Immunol.* **29**(12), 4037–42.
21. Lelouard, H. *et al.* (2002). *Nature* **417**(6885), 177–82.
22. Lelouard, H. *et al.* (2004). *J. Cell Biol.* **164**(5), 667–75.
23. Canadien, V. *et al.* (2005). *J. Immunol.* **174**(5), 2471–5.
24. Gil-Torregrosa, B. C. *et al.* (2004). *Eur. J. Immunol.* **34**(2), 398–407.
25. Ackerman, A. L. *et al.* (2003). *J. Immunol.* **170**(8), 4178–88.
26. Cella, M. *et al.* (1999). *J. Exp. Med.* **189**(5), 821–9.
27. Rescigno, M. *et al.* (1998). *Proc. Natl Acad. Sci. U S A* **95**(9), 5229–34.

28. Delamarre, L. *et al.* (2003). *J. Exp. Med.* **198**(1), 111–22.

29. Pamer, E. G. (2004). *Nat. Rev. Immunol.* **4**(10), 812–23.

30. Pamer, E. G. *et al.* (1997). *Immunol. Rev.* **158**, 129–36.

31. Villanueva, M. S. *et al.* (1994). *Immunity* **1**(6), 479–89.

32. Villanueva, M. S. *et al.* (1995). *J. Immunol.* **155**(11), 5227–33.

33. Princiotta, M. F. *et al.* (2003). *Immunity* **18**(3), 343–54.

34. Sijts, A. J. *et al.* (1997). *J. Biol. Chem.* **272**(31), 19261–8.

35. Shen, H. *et al.* (1998). *Cell* **92**(4), 535–45.

36. Nakagawa, I. *et al.* (2004). *Science* **306**(5698), 1037–40.

37. Ogawa, M. *et al.* (2005). *Science* **307**(5710), 727–31.

38. Rich, K. A. *et al.* (2003). *Cell Microbiol.* **5**(7), 455–68.

39. Portnoy, D. A. *et al.* (1989). *J. Exp. Med.* **170**(6), 2141–6.

40. Tvinnereim, A. R. *et al.* (2004). *J. Immunol.* **173**(3), 1994–2002.

41. Higgins, D. E. *et al.* (1999). *Mol. Microbiol.* **31**(6), 1631–41.

42. Hess, J. *et al.* (1998). *Proc. Natl Acad. Sci. U S A* **95**(9), 5299–304.

43. Russmann, H. *et al.* (1998). *Science* **281**(5376), 565–8.

44. Guermonprez, P. *et al.* (1999). *J. Immunol.* **162**(4), 1910–16.

45. Pfeifer, J. D. *et al.* (1993). *Nature* **361**(6410), 359–62.

46. Cheadle, E. J. *et al.* (2005). *Infect. Immun.* **73**(2), 784–94.

47. Winter, N. *et al.* (1995). *Vaccine* **13**(5), 471–8.

48. van Faassen, H. *et al.* (2004). *J. Immunol.* **172**(6), 3491–500.

49. Dudani, R. *et al.* (2002). *J. Immunol.* **168**(11), 5737–45.

50. Wick, M. J. *et al.* (1996). *Eur. J. Immunol.* **26**(11), 2790–9.

51. Shen, L. *et al.* (2004). *Immunity* **21**(2), 155–65.

52. Harding, C. V. *et al.* (1994). *J. Immunol.* **153**(11), 4925–33.

53. Song, R. *et al.* (1996). *J. Immunol.* **156**(11), 4182–90.

54. Chefalo, P. J. *et al.* (2003). *J. Immunol.* **170**(12), 5825–33.

55. Chefalo, P. J. *et al.* (2001). *J. Immunol.* **167**(3), 1274–82.

56. MacAry, P. A. *et al.* (2001). *Proc. Natl Acad. Sci. U S A* **98**(7), 3982–7.

57. Kleijmeer, M. J. *et al.* (2001). *Traffic* **2**(2), 124–37.

58. Turley, S. J. *et al.* (2000). *Science* **288**(5465), 522–7.

59. Lizee, G. *et al.* (2003). *Nat. Immunol.* **4**(11), 1065–73.

60. McCluskey, J. *et al.* (1986). *EMBO J.* **5**(10), 2477–83.

61. Vega, M. A. *et al.* (1989). *Proc. Natl Acad. Sci. U S A* **86**(8), 2688–92.

62. Kleijmeer, M. *et al.* (2001). *J. Cell Biol.* **155**(1), 53–63.

63. Chow, A. *et al.* (2002). *Nature* **418**(6901), 988–94.

64. Boes, M. *et al.* (2002). *Nature* **418**(6901), 983–8.

65. Svensson, M. *et al.* (1999). *Eur. J. Immunol.* **29**(1), 180–8.

66. Brumell, J. H. *et al.* (2002). *Curr. Biol.* **12**(1), R15–17.

67. Stamm, L. M. *et al.* (2003). *J. Exp. Med.* **198**(9), 1361–8.

68. McDonough, K. A. *et al.* (1993). *Infect. Immun.* **61**(7), 2763–73.
69. Roy, D. *et al.* (2004). *Science* **304**(5676), 1515–18.
70. Beuzon, C. R. *et al.* (2000). *EMBO. J.* **19**(13), 3235–49.
71. Teitelbaum, R. *et al.* (1999). *Proc. Natl Acad. Sci. U S A* **96**(26), 15190–5.
72. Kovacsovics-Bankowski, M. *et al.* (1995). *Science* **267**(5195), 243–6.
73. Reis e Sousa, C. *et al.* (1995). *J. Exp. Med.* **182**(3), 841–51.
74. Oh, Y. K. *et al.* (1997). *Vaccine* **15**(5), 511–18.
75. Norbury, C. C. *et al.* (1995). *Immunity* **3**(6), 783–91.
76. Norbury, C. C. *et al.* (1997). *Eur. J. Immunol.* **27**(1), 280–8.
77. Rodriguez, A. *et al.* (1999). *Nat. Cell Biol.* **1**(6), 362–8.
78. West, M. A. *et al.* (2004). *Science* **305**(5687), 1153–7.
79. Blander, J. M. *et al.* (2004). *Science* **304**(5673), 1014–18.
80. Shiratsuchi, A. *et al.* (2004). *J. Immunol.* **172**(4), 2039–47.
81. Gagnon, E. *et al.* (2002). *Cell* **110**(1), 119–31.
82. Garin, J. *et al.* (2001). *J. Cell Biol.* **152**, 165–80.
83. Guermonprez, P. *et al.* (2003). *Nature* **425**(6956), 397–402.
84. Ackerman, A. L. *et al.* (2003). *Proc. Natl Acad. Sci. USA* **100**(22), 12889–94.
85. Tsai, B. *et al.* (2002). *Nat. Rev. Mol. Cell Biol.* **3**(4), 246–55.
86. Tortorella, D. *et al.* (2000). *Annu. Rev. Immunol.* **18**, 861–926.
87. Wiertz, E. J. *et al.* (1996). *Nature* **384**(6608), 432–8.
88. Tirosh, B. *et al.* (2003). *J. Biol. Chem.* **278**(9), 6664–72.
89. Lilley, B. N. *et al.* (2004). *Nature* **429**(6994), 834–40.
90. Ye, Y. *et al.* (2004). *Nature* **429**(6994), 841–7.
91. Ye, Y. *et al.* (2001). *Nature* **414**(6864), 652–6.
92. Houde, M. *et al.* (2003). *Nature* **425**(6956), 402–6.
93. Ackerman, A. L. *et al.* (2005). *Nat. Immunol.* **6**(1), 107–13.
94. Roy, C. R. (2005). *Proc. Natl Acad. Sci. U S A* **102**(5), 1271–2.
95. Celli, J. *et al.* (2003). *J. Exp. Med.* **198**(4), 545–56.
96. Celli, J. *et al.* (2004). *Curr. Opin. Microbiol.* **7**(1), 93–7.
97. Nagai, H. *et al.* (2002). *Science* **295**(5555), 679–82.
98. Kagan, J. C. *et al.* (2004). *J. Exp. Med.* **199**(9), 1201–11.
99. McNew, J. A. *et al.* (2000). *Nature* **407**(6801), 153–9.
100. Becker, T. *et al.* (2005). *Proc. Natl Acad. Sci. U S A* **102**(11), 4022–6.
101. Trombetta, E. S. *et al.* (2005). *Annu. Rev. Immunol.* **23**, 975–1028.
102. Wilson, N. S. *et al.* (2005). *Adv. Immunol.* **86**, 241–305.
103. Lutz, M. B. *et al.* (1997). *J. Immunol.* **159**(8), 3707–16.
104. Pierre, P. *et al.* (1998). *Cell* **93**(7), 1135–45.
105. Trombetta, E. S. *et al.* (2003). *Science* **299**(5611), 1400–3.
106. Wilson, N. S. *et al.* (2004). *Blood* **103**(6), 2187–95.
107. Delamarre, L. *et al.* (2005). *Science* **307**(5715), 1630–4.

108. Ramachandra, L. *et al.* (1999). *J. Immunol.* **162**(6), 3263–72.
109. Ramachandra, L. *et al.* (2000). *J. Immunol.* **164**(10), 5103–12.
110. Ramachandra, L. *et al.* (2001). *J. Exp. Med.* **194**(10), 1421–32.
111. Geginat, G. *et al.* (2001). *J. Immunol.* **166**(3), 1877–84.
112. Skoberne, M. *et al.* (2002). *J. Immunol.* **168**(4), 1854–60.
113. Skoberne, M. *et al.* (2002). *J. Immunol.* **169**(3), 1410–18.
114. Tewari, M. K. *et al.* (2005). *Nat. Immunol.* **6**(3), 287–94.
115. Dani, A. *et al.* (2004). *J. Cell Sci.* **117**(Pt 18), 4219–30.
116. Cuervo, A. M. *et al.* (1998). *J. Mol. Med.* **76**(1), 6–12.
117. Zhou, D. *et al.* (2005). *Immunity* **22**(5), 571–81.
118. Paludan, C. *et al.* (2005). *Science* **307**(5709), 593–6.
119. Dengjel, J. *et al.* (2005). *Proc. Natl Acad. Sci. U S A* **102**(22), 7922–7.
120. Yoshimori, T. (2004). *Biochem. Biophys. Res. Commun.* **313**(2), 453–8.
121. Macpherson, A. J. *et al.* (2004). *Science* **303**(5664), 1662–5.
122. Macpherson, A. J. *et al.* (2000). *Science* **288**(5474), 2222–6.
123. Garcia-Del Portillo, F. *et al.* (2000). *Infect. Immun.* **68**(5), 2985–91.
124. Scott, C. C. *et al.* (2003). *J. Membr. Biol.* **193**(3), 137–52.
125. Sturgill-Koszycki, S. *et al.* (1994). *Science* **263**(5147), 678–81.
126. Vergne, I. *et al.* (2004). *Annu. Rev. Cell Dev. Biol.* **20**, 367–94.
127. Fratti, R. A. *et al.* (2001). *J. Cell Biol.* **154**(3), 631–44.
128. Fratti, R. A. *et al.* (2003). *Proc. Natl Acad. Sci. U S A* **100**(9), 5437–42.
129. Vergne, I. *et al.* (2003). *J. Exp. Med.* **198**(4), 653–9.
130. Vergne, I. *et al.* (2005). *Proc. Natl Acad. Sci. U S A* **102**(11), 4033–8.
131. Gordon, A. H. *et al.* (1994). *Infect. Immun.* **62**(10), 4650–1.
132. Harding, C. V. *et al.* (2003). *Curr. Opin. Immunol.* **15**(1), 112–19.
133. Ramachandra, L. *et al.* (2005). *Infect. Immun.* **73**(2), 1097–105.
134. Hmama, Z. *et al.* (1998). *J. Immunol.* **161**(9), 4882–93.
135. Chan, J. *et al.* (2004). *Clin. Immunol.* **110**(1), 2–12.
136. Jantsch, J. *et al.* (2003). *Cell Microbiol.* **5**(12), 933–45.
137. Cheminay, C. *et al.* (2005). *J. Immunol.* **174**(5), 2892–9.
138. Tobar, J. A. *et al.* (2004). *J. Immunol.* **173**(6), 4058–65.
139. Tailleux, L. *et al.* (2003). *J. Immunol.* **170**(4), 1939–48.
140. Neild, A. L. *et al.* (2003). *Immunity* **18**(6), 813–23.
141. Bevan, M. J. (1976). *J. Exp. Med.* **143**(5), 1283–8.
142. Heath, W. R. *et al.* (2004). *Immunol. Rev.* **199**, 9–26.
143. Yewdell, J. W. *et al.* (2005). *Annu. Rev. Immunol.* **23**, 651–82.
144. Ramirez, M. C. *et al.* (2002). *J. Immunol.* **169**(12), 6733–42.
145. Fonteneau, J. F. *et al.* (2003). *Blood* **102**(13), 4448–55.
146. Guermonprez, P. *et al.* (2005). *Springer Semin. Immunopathol.* **26**(3), 257–71.

147. Shen, L. *et al.* (2004). *Proc. Natl Acad. Sci. U S A* **101**(9), 3035–40.
148. Wolkers, M. C. *et al.* (2004). *Science* **304**(5675), 1314–17.
149. Norbury, C. C. *et al.* (2004). *Science* **304**(5675), 1318–21.
150. Blachere, N. E. *et al.* (2005). *PLoS Biol.* **3**(6), e185.
151. Binder, R. J. *et al.* (2005). *Nat. Immunol.* **6**(6), 593–9.
152. Srivastava, P. (2002). *Nat. Rev. Immunol.* **2**(3), 185–94.
153. Tobian, A. A. *et al.* (2004). *J. Immunol.* **172**(9), 5277–86.
154. Schaible, U. E. *et al.* (2003). *Nat. Med.* **9**(8), 1039–46.
155. Wijburg, O. L. *et al.* (2002). *J. Immunol.* **169**(6), 3275–83.
156. Wong, P. *et al.* (2003). *Immunity* **18**(4), 499–511.
157. Janda, J. *et al.* (2004). *J. Immunol.* **173**(9), 5644–51.

PART II Dendritic cells and innate immune responses to bacteria

CHAPTER 4

Dendritic cell activation and uptake of bacteria in vivo

Maria Rescigno
European Institute of Oncology

4.1 INTRODUCTION

Pathogenic bacteria have evolved several strategies to gain access across epithelial surfaces particularly those lining the mucosae. After their epithelial transcytosis bacteria find a first line of immune defense represented by professional phagocytes, including macrophages and dendritic cells. These cells are particularly apt at bacterial uptake, killing and processing for the initiation/maintenance of adaptive immune responses. Furthermore, intracellular bacteria can induce by epithelial cells the release of inflammatory mediators and cytokines that will recruit other immune cells, particularly neutrophils. Dendritic cells are not simply passive players waiting for possible invaders, they can actively participate to bacterial sampling by intercalating between epithelial cells. This mechanism is not restricted to pathogenic bacteria. Since gut dendritic cells have been thoroughly studied, in this chapter we will focus on dendritic cells located in the intestinal mucosa and on their role in the uptake and handling of luminal bacteria.

4.2 THE ANATOMY OF THE INTESTINAL MUCOSAL EPITHELIUM AND THE GUT ASSOCIATED LYMPHOID TISSUE (GALT)

The intestinal epithelium is the first line of defense toward dangerous microorganisms[1,2]. It opposes a physical, electric and chemical barrier against luminal bacteria. The permeability of the barrier is regulated by the presence of both tight junctions (TJ) between epithelial cells (ECs) and a negatively charged mucous glycocalix. TJ seal adjacent ECs to one another and regulate solute and ion flux between cells[3]. The glycocalix sets the size of macromolecules that can reach the apical membrane of ECs [4] and opposes an electric barrier to bacteria. Finally, ECs and Paneth cells, specialized cells located at the base of the crypt of intestinal villi, release antimicrobial peptides including defensins and cathelicidins that target broad classes of microorganisms[5]. The intestinal epithelial barrier is further complicated by the presence of two important cell types that are interspersed between ECs and play a crucial role in sampling the luminal content: (microfold) M cells[6] and DCs[7–9]. M cells are found primarily in the follicle-associated epithelium (FAE) of Peyer's patches (PP) but they have been recently described to be scattered also among the absorptive epithelium where they could potentially transport antigens to the lamina propria (LP)[10]. M cells, differently from ECs, do not have an organized brush border and are more permissive to antigen uptake[4]. DCs are phagocytic cells that are scattered throughout the intestinal epithelium[11]. It has been recently described that DCs are able to send dendrites out like periscopes into the lumen for bacterial uptake[12,13]. The integrity of the epithelial barrier is preserved because DCs express TJ proteins and can establish new TJ-like structure with adjacent ECs[12]. These "creeping" DCs are characterized by the expression of the myeloid marker CD11b and the lack of CD8α[13,14]. Their presence in the terminal ileum where the gradient of bacteria gradually increases suggests they may be recruited by the presence of luminal bacteria. Interestingly, DCs in CX3CL1 (fractalkine) receptor-deficient mice are unable to spread their dendrites across the epithelial barrier, indicating the involvement of CX3CL1 in driving the extension of the dendrites[14]. It is unknown whether bacteria can directly drive fractalkine production by epithelial cells and whether fractalkine modulates TJ protein expression in DCs. Interestingly, bacteria lacking LPS are unable to recruit DCs in in vitro generated epithelial cell monolayers suggesting that bacteria play an active role in the induction of DC migration across the epithelial barrier[15].

The GALT can be divided into inductive sites where the immune response is initiated and effector sites where immune cells carry out their function[2,16]. Peyer's patches, mesenteric lymph nodes (MLN) and isolated lymphoid follicles are important inductive sites for mucosal immune responses whereas the epithelium and the lamina propria of the mucosa are considered effector sites for antibody production and T cell responses.

4.3 BACTERIAL UPTAKE IN THE GUT AND MUCOSAL DC SUBPOPULATIONS

In the absence of concomitant activation stimuli, LP-DCs are probably involved in the induction of oral tolerance. In fact expansion of DCs in vivo enhanced tolerance induction after antigen feeding[17]. It is possible that antigen-loaded DCs migrate to MLN which is the preferential site for naive T cell activation and expansion after oral feeding of soluble antigen[18]. Conversely, particulate antigen is most likely taken up in PP as mice lacking PP are perfectly competent to induce antibody response toward soluble but not toward particulate (microsphere) antigen[19].

The mechanisms of bacterial entrance depend on their pathogenicity (Figure 4.1). Most of the pathogens have developed strategies to penetrate ECs or to facilitate M cell invasion (for a review see Ref. 1), whereas non-invasive bacteria can enter mucosal surfaces either through M cells or DCs. M cells can release their "cargo" to underlying phagocytic cells, including DCs, that can migrate to the interfollicular region of PP for T and B cell interactions, whereas DCs that take up bacteria directly across mucosal surfaces are likely to migrate to MLN. Interestingly, MLN set the border for mucosal compartment avoiding systemic spread of commensal-loaded DCs[20]. Both mechanisms do not discriminate between invasive pathogenic and non-invasive commensal bacteria. An alternative mechanism for antigen entry across a mucosal surface that also targets DCs and could be used for bacterial internalization, has been recently described[21]. It is mediated by neonatal Fc receptors (FcRn) expressed by adult human (but not mouse) intestinal epithelial cells that transport IgG across the intestinal epithelial barrier, and after binding with cognate antigen in the intestinal lumen, recycles the immune complexes back to the LP[21]. Antigens bound by IgG are less susceptible to degradation within the epithelial cells because endosomes formed after uptake by FcRn do not readily fuse with lysosomes. FcRn transport directs and delivers the antigens in the form of immune complexes directly to DCs lying in the LP. As DCs can be activated by immune complexes, it would be interesting to know whether DCs internal-ize the immune complexes via the FcγRs or via FcRn (both of which are expressed by DCs) and whether these receptors differentially affect DC function. The role of IgA and their secretory form (sIgA) in facilitating the internalization of opsonized bacteria still needs to be investigated but it is known that IgA-coated antigens although being excluded from epithelial cell binding, are facilitated in their access across M cells[22] and it has been recently shown that they are targeted directly to DCs present in the dome region of PPs[23]. Finally, DCs can process antigens from apoptotic intestinal epithelial cells, both in the steady state[24] and following reovirus infection[25], which constitutes another mechanism of DC antigen uptake that directly involves interactions with the epithelium.

Pathways of bacterial entry

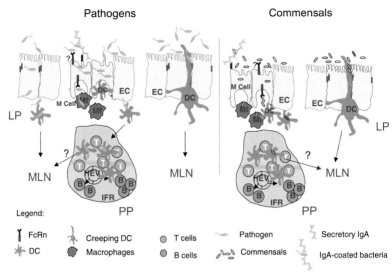

Figure 4.1. Mechanisms of bacterial uptake. The mechanisms of bacterial entrance depend on their pathogenicity. Most of the pathogens have developed strategies to penetrate ECs or to facilitate M cell invasion via the expression of type three secretion system and invasive genes; alternatively they are captured by creeping DCs (left). Commensal bacteria can enter mucosal surfaces either through M cells or DCs (right). It is not yet clear whether entrance through M cells of commensal bacteria is restricted to immunoglobulin-coated bacteria. In this case, internalization of IgG-coated bacteria could be mediated by neonatal Fc receptors (FcRn) expressed by adult human (but not mouse) intestinal epithelial cells, whereas secretory (s)IgA-coated bacteria could be internalized by IgA receptor expressed by M cells. M cells can release their "cargo" to underlying phagocytic cells, including DCs, that can migrate to the interfollicular region (IFR) of Peyer's Patches for T and B cell interactions, whereas DCs that take up bacteria directly across mucosal surfaces are likely to migrate to MLN. Alternatively, PP-DCs could migrate to MLN. HEV: high endothelial venules. (For a colour version of this figure, please refer to the colour insert between pages 12 and 13.)

The uptake route together with the nature of the ingested antigens dictates the type of immune response that is generated, whether this is related to the subtype of DCs that is targeted by each route or to their location remains to be established. In fact at least four DC populations in the mouse intestine have been described. They are all characterized by the expression of CD11c but differ for the expression of the surface markers CD11b, CD8α and B220 (for a review see Refs 8 and 26) as well as for the expression of chemokine receptors CCR6 and CCR7[27].

Interestingly, the different DC populations have particular locations in PP[28]. In fact, it is important to say that in PP two important functions are carried out by DCs: uptake of antigen after its transcytosis across the FAE and T and B cell activation. Therefore, differently from other peripheral tissues, it is possible to find in the PP both immature DCs that are mainly localized in the sub-epithelial dome, below the FAE and mature DCs that are found in interfollicular T cell areas. Two additional DC subsets have been described in MLN that are characterized by the differential expression of CD4 and DEC-205[8,26]. The characterization of human intestinal DCs is still very poor, but at least two DC types have been described in the colon: a CD11c[+]HLA-DR[+] population and a CD11c[−] population[29] that we have identified as CD83[+]CD123[+], possibly plasmacytoid DCs (our unpublished observations). Hence, scattered throughout mucosal tissues it is possible to find the same DC subsets present in other non-mucosal tissues (see Chapter 1).

4.4 HANDLING OF BACTERIA BY EPITHELIAL CELLS CAN INFLUENCE THE INDUCTION OF IMMUNE RESPONSES

The major interaction between mucosal tissues and luminal bacteria occurs at the level of ECs that are the most representative cell type of the epithelium. Both pathogens and commensal bacteria have been described to undertake an active cross-talk with ECs[1]. Whereas the first are primarily involved in the activation of an inflammatory cascade of events, the latter seem to downregulate the ability of ECs to initiate inflammatory responses. The mechanisms through which pathogens can activate ECs are similar to those used by monocytes and DCs to sense the presence of bacteria. In fact ECs express a series of pathogen recognition receptors (PRRs, see Chapters 2 and 9) including Toll-like receptors (TLRs) and NOD proteins that are expressed also by phagocytes[1]. The major difference stands in the location of these receptors. In fact ECs seem to express these receptors either intracellularly (like TLR-4) or in a polarized fashion leaving the apical surface nearly free of PRR expression. Therefore only invasive bacteria or those equipped with type three or four secretion systems[30] that act as syringes to pump DNA or effector proteins directly into the cytoplasm of host cells, are sensed by PRRs for activation of the inflammatory cascade. Moreover, some of the receptors (like NOD2) are constitutively expressed only in Paneth cells[31] that reside at the base of the cripts and are induced in ECs only after bacterial encounter[32,33]. A typical indicator of epithelial infection by invasive bacteria is the expression of the chemokine CXCL-8 (IL-8) which is a strong chemoattractant for neutrophils[34-37]. A more debated issue relates to the expression of TLR-5, the receptor for flagellin[38]. It has been recently described that flagellin-dependent stimulation of intestinal ECs results in triggering of CCL20 via a TLR-5 dependent mechanism[39]. CCL20 is responsible for the recruitment of CCR6-expressing immature DCs[40]. However, some authors suggest that TLR-5 is expressed only basolaterally of ECs[41,42], whereas others have also described it apically[39,43]. We favor the second hypothesis because we have evidence that invasive-deficient mutant of *Salmonella* and the flagellated non-invasive soil bacterium *Bacillus subtilis* induce the expression of CCL-20 by polarized ECs[15,44]. Our experiments in the mouse also confirm that non-invasive flagellated bacteria can induce the expression of CCL-20 suggesting the possibility that different responses might depend on the EC cell line used for in vitro experiments[44].

How commensals can downregulate the inflammatory response induced by pathogen associated molecular patterns (PAMPs) has only recently started to be unraveled. It is becoming clear that recognition of commensal flora via TLRs is required for intestinal homeostasis[45] and that commensal bacteria can interfere at different levels of TLR signaling. Expression and activation of IRAK-M[46] or of a truncated version of the TLR adaptor protein MyD88[47] that both interfere with TLR signaling have been described. Along the same line, the interaction of ECs with the commensal *Bacteroides thetaiotamicron* or with non-virulent mutants of *Salmonella typhimurium* interfere with the activation of NF-κB that is downstream of TLR signaling either by triggering binding of peroxisome-proliferator-activated receptor γ (PPAR-γ) with the NF-κB subunit Rel-A in the nucleus[48] or by blocking the degradation of IκBα, an intracellular inhibitor of NF-κB[49]. Therefore, the induction of an inflammatory response in ECs depends on the ability of invasive pathogens to activate PRR signaling pathways and on that of commensals to perturb the same signaling pathways.

4.5 UNIQUE FUNCTIONS OF MUCOSAL DCs

DCs isolated from a variety of mucosal sites (PP, LP, mesenteric lymph nodes (MLN), lung) have the natural propensity to induce T_H2 responses in in vitro T cell priming assays, and to express cytokines such as IL-10, and possibly TGF-β[2,50-53]. Interestingly, the same CD11c⁺CD11b⁺CD8α⁻ DC subset isolated from PP but not from spleen preferentially polarizes antigen-specific T cells to produce T_H2 cytokines and IL-10 in vitro[54], suggesting that the observed differences are not attributable to subset-intrinsic properties but most likely to the local mucosal microenvironment. Further, the same PP but not spleen DC subset is able to promote IgA production by naive B cells, which is mediated by a higher release of IL-6[55] and T cell help. These data suggest that mucosal DCs may be specialized in inducing a non-inflammatory environment and in providing help to B cells via the activation of T_H2 T cells. This is consistent with the fact that many "tolerogenic" responses to mucosal antigens, for example to commensal organisms, are associated with the generation of antibody responses[20,51], rather than with a broad immunological unresponsiveness. In addition, $CD8^+$ $CD11c^{lo}$ plasmacytoid DCs may also be important for maintaining tolerance to innocuous antigens since this population can induce the differentiation of IL-10 producing regulatory T cells (Treg) in vitro[8].

Another important feature of DCs isolated from mucosal tissues is that they have the unique ability to selectively imprint gut-homing T cells[56-58]. Moreover, naive $CD8^+$ T cells primed by PP-DCs acquire gut tropism[57], despite showing similar patterns of activation markers and effector activity as those primed by DCs isolated from other non-mucosal lymphoid organs. PP-DCs induced high expression of the intestinal homing integrin $\alpha_4\beta_7$ and the chemokine receptor CCR9 in primed $CD8^+$ T cells. Interestingly, reactivation of skin-committed memory T cells with DCs isolated from gut changed T cell tissue tropism, suggesting that memory T cells are relocated according to the tissue where they are needed[59]. Finally, mucosal DCs have been shown to continually migrate to draining lymph nodes in the "steady", or unperturbed-state with a rapid turnover rate (2−4 days in the intestinal wall). In the rat, two types of migrating DCs could be identified, both of which are positive for the αE integrin CD103, but only the fraction that expresses low levels of CD172 (SIRPα), has features of immature cells and carries apoptotic enterocytes to MLNs[60]. Because these DCs process apoptotic epithelial cells in the steady-state[24], this $CD103^+CD172^{lo}$ DC population may be involved in tolerance to self-proteins, although this hypothesis remains to be tested. DC emigration from the gut can be greatly enhanced by systemic LPS injection which does not change the proportion of SIRPα^{hi}/SIRPα^{lo} populations as well as their activation state[61]. Interestingly, whereas SIRPα^{lo} DCs migrate to T cell areas of MLNs under steady-state conditions, SIRPα^{hi} DCs do so only after intravenous LPS injection suggesting that LPS injection facilitates antigen presentation by this DC subset.

4.6 THE CROSS-TALK BETWEEN ECs AND DCs HELPS REGULATING THE INDUCTION OF MUCOSAL IMMUNE RESPONSES

DCs play an active role in bacterial uptake across mucosal surfaces and have unique functions that allow the generation of mucosal immune responses. Moreover, DCs can intercalate between ECs and can interact directly with the luminal bacteria and with all the TLR ligands that are carried by commensal or pathogenic bacteria. Therefore, three important questions arise: what is the role played by the local microenvironment in driving mucosal DC differentiation? How can DCs avoid the induction of exaggerated inflammatory responses toward commensal bacteria? Is there any relationship between the unique phenotype of mucosal DCs and the regulation of gut immune homeostasis? One possibility is that DCs sense differently commensal versus pathogenic bacteria. However, rather than a difference between dangerous and non-dangerous bacteria it seems that Gram negative versus Gram positive bacteria are differentially sensed by DCs. In fact, Gram negative bacteria (including commensals) activate DCs that induce a Th1 type of response, by contrast, Gram positive bacteria are more prone to induce Th2 type of responses[62]. Certain probiotics belonging to the *Lactobacillus* species (*L. reuteri* and *L. casei*, but not *L. plantarum*) can drive tolerogenic DCs[63]. Therefore, it is not possible to generalize on the possibility of DCs to sense differently commensals, but some species could participate in downmodulating DC function. It is becoming clear that the relationship between DCs and the microenvironment are profoundly affecting the functional properties of tissue DCs. This has been demonstrated in the spleen[64,65], but there is strong evidence that a similar situation is occurring also in the gut. In fact, the ability of intestinal DCs to induce gut-tropism during T cell priming[56−58] and reactivation[59] and to promote T_H2 T cell responses[2,50−54], as well as IgA antibody production[55] strongly favors this hypothesis. As intestinal ECs are in close contact with DCs, they could play an active role in driving mucosal DC differentiation. This is indeed the case because ECs release constitutively thymic stromal lymphopoietin (TSLP), a molecule involved in driving T_H2 differentiation by DCs[66,67]. Interestingly, DCs exposed to EC-conditioning are unable to release IL-12 and to drive T_H1 type of T cell responses even after activation with T_H1-inducing pathogens (Figure 4.2)[68]. Moreover, TSLP acts in a very narrow window of concentrations: at lower or higher TSLP concentrations, DCs reacquire the ability to release IL-12 and to drive T_H1 T cell responses. Therefore, it is likely that resident DCs even though they have the chance to

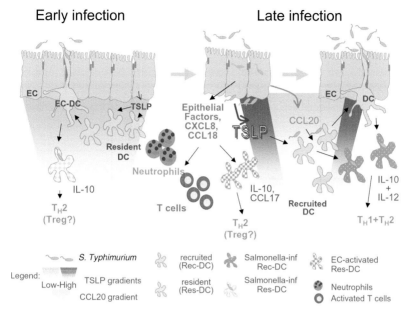

Figure 4.2. The cross-talk between ECs and DCs in *Salmonella typhimurium* handling. Early *Salmonella typhimurium* infection: resident DCs are conditioned by EC-released TSLP (Res-DC). Res-DC release IL-10 after bacterial exposure and drive default T_H2 responses to *S. typhimurium*. It remains to be established whether these cells also drive T regulatory cells. Late infection: since *Salmonella typhimurium* is an invasive bacterium, it induces ECs to release pro-inflammatory chemokines like IL-8 (CXCL-8) and PARC (CCL-18), which attract neutrophils, granulocytes and activated T cells that generate an inflamed site. The binding of *Salmonella* to the basolateral membrane of ECs induces the upregulation of TSLP. TSLP at this concentration drives T_H1 rather than T_H2 promoting DCs in response to bacteria. Unidentified EC-derived factors can also activate "bystander" DCs that have not been in contact directly with the bacteria. DCs activated in this way release IL-10 and TARC (CCL-17) but not IL-12, thus driving and recruiting T_H2 T cells and maybe T regulatory cells. *Salmonella* also induces the release of MIP-3α (CCL-20) that recruits CCR6-expressing immature DCs. Most likely, recruited DCs are not subjected to EC-conditioning, rather they could find increased TSLP concentrations in the infected site. Newly recruited DCs (Rec-DC) can either creep between ECs to take up bacteria or they can phagocytose bacteria that have breached across the epithelial barrier and release both IL-10 and IL-12, thus promoting T_H1 and T_H2 responses. This allows the establishment of protective anti-*Salmonella* responses. (For a colour version of this figure, please refer to the colour insert between pages 12 and 13.)

contact directly the bacteria, they are unable to activate inflammatory cells and this can help to maintain the homeostasis of the gut. In fact, nearly 70 per cent of individuals affected by a T_H1-mediated chronic inflammatory disease like in Crohn's disease[69] have undetectable levels of TSLP and this correlates with the inability of intestinal ECs to regulate DC function[68]. Therefore, resident DCs that are actively involved in taking up bacteria at steady-state do not drive inflammatory responses and this can explain why the intestinal immune homeostasis is preserved even though DCs are continuously exposed to TLR ligands.

Despite this propensity for the induction of T_H2 and Tregs by mucosal DCs, T_H1 and cytotoxic T lymphocyte (CTL) responses are effectively generated to mucosal pathogens and are required to fight intracellular microorganisms[70–74]. Whether this involves the same or different DC subsets as those responsible for mucosal responses and tolerance induction, remains to be established. However, it is conceivable that resident mucosal DCs are "educated" by ECs to initiate non-inflammatory responses, whereas DCs recruited after bacterial invasion might retain their ability to respond in an inflammatory mode. In fact, infection by flagellated bacteria like *Salmonella* spp. induces the recruitment of DCs in the intestinal epithelium[12,14] via the release of CCL-20 by ECs[39]. These non-conditioned newly recruited DCs might be responsible for the induction of T_H1 responses to invasive bacteria (Figure 4.2). This hypothesis is supported by in vitro three-partite studies in which DCs were seeded from the basolateral membrane of EC monolayers shortly before apical bacterial application[68]. Interestingly, due to their ability to creep between ECs and to contact bacteria directly, DCs were "qualitatively" similarly activated regardless of the invasiveness or pathogenicity of the apical bacteria. Bacteria-activated DCs produced both IL-12 and IL-10 and skewed toward a T_H1 phenotype[68]. This suggests that non-conditioned DCs can drive the induction of inflammatory responses provided that they are not subject to EC conditioning before their encounter with bacteria. Moreover, bacteria invading ECs induce the upregulation of TSLP thus switching to DCs that have the propensity to induce T_H1- rather than T_H2-T cells in response to bacteria. Interestingly, bystander DCs that do not contact directly the bacteria are activated by EC-derived factors to non-inflammatory DCs producing IL-10 and TARC (CCL-17) and inducing or recruiting TH2 T cells, probably as a feedback mechanism to turn off the inflammatory response[44]. Whether these DCs also drive T regulatory cells has to be investigated.

Another possibility is that epithelial cell derived factors, such as TNF or type 1 IFNs, produced during pathogen invasion may directly affect

DC activation. This hypothesis is supported by studies of murine intestinal infection with type-1 reovirus[25]. Reovirus productively infects epithelial cells overlying PPs, yet viral antigen associated with apoptotic epithelial cells is avidly taken up by CD11c$^+$ CD8α^- CD11b$^-$ DCs in the subepithelial dome[25]. The observation that reovirus neither productively infects DCs in vivo or in vitro, nor activates DCs to mature or produce cytokines in vitro, suggests a role for environmental factors, possibly derived from infected epithelial cells, in driving DCs to induce T$_H$1 responses to the virus. Interestingly, IFNαβR-deficient mice, but not MyD88-deficient or TLR3-deficient mice have an increased susceptibility to reovirus infection. In addition, MyD88-deficient mice mount normal IgG1, IgG2a/c and IgG2b responses, suggesting that type 1 IFN, possibly derived in the early stages of infection from infected epithelial cells, but not signaling via at least a single TLR pathway is important for inducing protection from reovirus infection.

Taken together, these studies highlight an important emerging relationship between DCs and epithelial cells in the maintenance of mucosal homeostasis and the induction of innate and adaptive immunity to mucosal infection with pathogens.

4.7 CONCLUSIONS

In conclusion, the outcome of bacterial handling by DCs at mucosal surfaces depends on several factors. These include: (a) the ability of DCs to discriminate between different sorts of bacteria through differential TLR engagement; (b) the unique specialized functions of mucosal DCs that allow the establishment of mucosal immune responses, including the induction of T$_H$2 T cell responses and IgA antibody production; (c) the interplay between epithelial cells and DCs at steady-state and during infection. Therefore, DCs play a crucial role both in the uptake of intestinal bacteria and in the induction of tolerance and immunity towards them. However, it remains to be clarified whether different DC subsets have clearly distinct functions in vivo or whether the local microenvironment is responsible to control DC function.

REFERENCES

1. Sansonetti, P. J. (2004). War and peace at mucosal surfaces. *Nat. Rev. Immunol.* **4**, 953–64.
2. Mowat, A. M. (2003). Anatomical basis of tolerance and immunity to intestinal antigens. *Nat. Rev. Immunol.* **3**, 331–41.

3. Schneeberger, E. E. and Lynch, R. D. (2004). The tight junction: a multifunctional complex. *Am. J. Physiol. Cell Physiol.* **286**, C1213–28.

4. Frey, A., Giannasca, K. T., Weltzin, R., Giannasca, P. J., Reggio, H., Lencer, W. I. and Neutra, M. R. (1996). Role of the glycocalyx in regulating access of microparticles to apical plasma membranes of intestinal epithelial cells: implications for microbial attachment and oral vaccine targeting. *J. Exp. Med.* **184**, 1045–59.

5. Ganz, T. (2003). Defensins: antimicrobial peptides of innate immunity. *Nat. Rev. Immunol.* **3**, 710–20.

6. Kraehenbuhl, J. P. and Neutra, M. R. (2000). Epithelial M cells: differentiation and function. *Annu. Rev. Cell Dev. Biol.* **16**, 301–32.

7. Banchereau, J., Briere, F., Caux, C., Davoust, J., Lebecque, S., Liu, Y. J., Pulendran, B. and Palucka, K. (2000). Immunobiology of dendritic cells. *Annu. Rev. Immunol.* **18**, 767–811.

8. Bilsborough, J. and Viney, J. L. (2004). Gastrointestinal dendritic cells play a role in immunity, tolerance, and disease. *Gastroenterology* **127**, 300–9.

9. Kelsall, B. L. and Rescigno, M. (2004). Mucosal dendritic cells in immunity and inflammation. *Nat. Immunol.* **5**, 1091–5.

10. Jang, M. H., Kweon, M. N., Iwatani, K., Yamamoto, M., Terahara, K., Sasakawa, C., Suzuki, T., Nochi, T., Yokota, Y., Rennert, P. D. *et al.* (2004). Intestinal villous M cells: an antigen entry site in the mucosal epithelium. *Proc. Natl Acad. Sci. U S A* **101**, 6110–15.

11. Maric, I., Holt, P. G., Perdue, M. H. and Bienenstock, J. (1996). Class II MHC antigen (Ia)-bearing dendritic cells in the epithelium of the rat intestine. *J. Immunol.* **156**, 1408–14.

12. Rescigno, M., Urbano, M., Valzasina, B., Francolini, M., Rotta, G., Bonasio, R., Granucci, F., Kraehenbuhl, J. P. and Ricciardi-Castagnoli, P. (2001). Dendritic cells express tight junction proteins and penetrate gut epithelial monolayers to sample bacteria. *Nat. Immunol.* **2**, 361–7.

13. Rescigno, M., Rotta, G., Valzasina, B. and Ricciardi-Castagnoli, P. (2001). Dendritic cells shuttle microbes across gut epithelial monolayers. *Immunobiology* **204**, 572–81.

14. Niess, J. H., Brand, S., Gu, X., Landsman, L., Jung, S., McCormick, B. A., Vyas, J. M., Boes, M., Ploegh, H. L., Fox, J. G. *et al.* (2005). CX3CR1-mediated dendritic cell access to the intestinal lumen and bacterial clearance. *Science* **307**, 254–8.

15. Rimoldi, M., Chieppa, M., Vulcano, M., Allavena, P. and Rescigno, M. (2004). Intestinal epithelial cells control DC function. *Ann. N Y Acad. Sci.* **1029**, 1–9.

16. Nagler-Anderson, C. (2001). Man the barrier! Strategic defences in the intestinal mucosa. *Nat. Rev. Immunol.* **1**, 59−67.

17. Viney, J. L., Mowat, A. M., O'Malley, J. M., Williamson, E. and Fanger, N. A. (1998). Expanding dendritic cells in vivo enhances the induction of oral tolerance. *J. Immunol.* **160**, 5815−25.

18. Kunkel, D., Kirchhoff, D., Nishikawa, S., Radbruch, A. and Scheffold, A. (2003). Visualization of peptide presentation following oral application of antigen in normal and Peyer's patches-deficient mice. *Eur. J. Immunol.* **33**, 1292−301.

19. Kunisawa, J., Takahashi, I., Okudaira, A., Hiroi, T., Katayama, K., Ariyama, T., Tsutsumi, Y., Nakagawa, S., Kiyono, H. and Mayumi, T. (2002). Lack of antigen-specific immune responses in anti-IL-7 receptor alpha chain antibody-treated Peyer's patch-null mice following intestinal immunization with microencapsulated antigen. *Eur. J. Immunol.* **32**, 2347−55.

20. Macpherson, A. J. and Uhr, T. (2004). Induction of protective IgA by intestinal dendritic cells carrying commensal bacteria. *Science* **303**, 1662−5.

21. Yoshida, M., Claypool, S. M., Wagner, J. S., Mizoguchi, E., Mizoguchi, A., Roopenian, D. C., Lencer, W. I. and Blumberg, R. S. (2004). Human neonatal Fc receptor mediates transport of IgG into luminal secretions for delivery of antigens to mucosal dendritic cells. *Immunity* **20**, 769−83.

22. Weltzin, R., Lucia-Jandris, P., Michetti, P., Fields, B. N., Kraehenbuhl, J. P., and Neutra, M. R. (1989). Binding and transepithelial transport of immunoglobulins by intestinal M cells: demonstration using monoclonal IgA antibodies against enteric viral proteins. *J. Cell Biol.* **108**, 1673−85.

23. Rey, J., Garin, N., Spertini, F. and Corthesy, B. (2004). Targeting of secretory IgA to Peyer's patch dendritic and T cells after transport by intestinal M cells. *J. Immunol.* **172**, 3026−33.

24. Huang, F. P., Platt, N., Wykes, M., Major, J. R., Powell, T. J., Jenkins, C. D. and MacPherson, G. G. (2000). A discrete subpopulation of dendritic cells transports apoptotic intestinal epithelial cells to T cell areas of mesenteric lymph nodes, see comments. *J. Exp. Med.* **191**, 435−44.

25. Fleeton, M. N., Contractor, N., Leon, F., Wetzel, J. D., Dermody, T. S. and Kelsall, B. L. (2004). Peyer's patch dendritic cells process viral antigen from apoptotic epithelial cells in the intestine of reovirus-infected mice. *J. Exp. Med.* **200**, 235−45.

26. Shortman, K. and Liu, Y. J. (2002). Mouse and human dendritic cell subtypes. *Nat. Rev. Immunol.* **2**, 151−61.

27. Iwasaki, A. and Kelsall, B. L. (2000). Localization of distinct Peyer's patch dendritic cell subsets and their recruitment by chemokines macrophage inflammatory protein (MIP)-3alpha, MIP-3beta, and secondary lymphoid organ chemokine. *J. Exp. Med.* **191**, 1381–94.

28. Niedergang, F., Didierlaurent, A., Kraehenbuhl, J. P. and Sirard, J. C. (2004). Dendritic cells: the host Achille's heel for mucosal pathogens? *Trends Microbiol.* **12**, 79–88.

29. Bell, S. J., Rigby, R., English, N., Mann, S. D., Knight, S. C., Kamm, M. A. and Stagg, A. J. (2001). Migration and maturation of human colonic dendritic cells. *J. Immunol.* **166**, 4958–67.

30. Viala, J., Chaput, C., Boneca, I. G., Cardona, A., Girardin, S. E., Moran, A. P., Athman, R., Memet, S., Huerre, M. R., Coyle, A. J. *et al.* (2004). Nod1 responds to peptidoglycan delivered by the *Helicobacter pylori* cag pathogenicity island. *Nat. Immunol.* **5**, 1166–74.

31. Lala, S., Ogura, Y., Osborne, C., Hor, S. Y., Bromfield, A., Davies, S., Ogunbiyi, O., Nunez, G. and Keshav, S. (2003). Crohn's disease and the NOD2 gene: a role for paneth cells. *Gastroenterology* **125**, 47–57.

32. Rosenstiel, P., Fantini, M., Brautigam, K., Kuhbacher, T., Waetzig, G. H., Seegert, D. and Schreiber, S. (2003). TNF-alpha and IFN-gamma regulate the expression of the NOD2 (CARD15) gene in human intestinal epithelial cells. *Gastroenterology* **124**, 1001–9.

33. Gutierrez, O., Pipaon, C., Inohara, N., Fontalba, A., Ogura, Y., Prosper, F., Nunez, G. and Fernandez-Luna, J. L. (2002). Induction of Nod2 in myelomonocytic and intestinal epithelial cells via nuclear factor-kappa B activation. *J. Biol. Chem.* **277**, 41701–5.

34. Eckmann, L., Kagnoff, M. F. and Fierer, J. (1993). Epithelial cells secrete the chemokine interleukin-8 in response to bacterial entry. *Infect. Immun.* **61**, 4569–74.

35. Jung, H. C., Eckmann, L., Yang, S. K., Panja, A., Fierer, J., Morzycka-Wroblewska, E. and Kagnoff, M. F. (1995). A distinct array of proinflammatory cytokines is expressed in human colon epithelial cells in response to bacterial invasion. *J. Clin. Invest.* **95**, 55–65.

36. McCormick, B. A., Colgan, S. P., Delp-Archer, C., Miller, S. I. and Madara, J. L. (1993). *Salmonella typhimurium* attachment to human intestinal epithelial monolayers: transcellular signalling to subepithelial neutrophils. *J. Cell Biol.* **123**, 895–907.

37. McCormick, B. A., Hofman, P. M., Kim, J., Carnes, D. K., Miller, S. I. and Madara, J. L. (1995). Surface attachment of *Salmonella typhimurium* to intestinal epithelia imprints the subepithelial matrix with gradients chemotactic for neutrophils. *J. Cell Biol.* **131**, 1599–608.

38. Hayashi, F., Smith, K. D., Ozinsky, A., Hawn, T. R., Yi, E. C., Goodlett, D. R., Eng, J. K., Akira, S., Underhill, D. M. and Aderem, A. (2001). The innate immune response to bacterial flagellin is mediated by Toll-like receptor 5. *Nature* **410**, 1099–103.

39. Sierro, F., Dubois, B., Coste, A., Kaiserlian, D., Kraehenbuhl, J. P. and Sirard, J. C. (2001). Flagellin stimulation of intestinal epithelial cells triggers CCL20-mediated migration of dendritic cells. *Proc. Natl. Acad. Sci. U S A* **98**, 13722–7.

40. Sozzani, S., Allavena, P., D'Amico, G., Luini, W., Bianchi, G., Kataura, M., Imai, T., Yoshie, O., Bonecchi, R. and Mantovani, A. (1998). Differential regulation of chemokine receptors during dendritic cell maturation: a model for their trafficking properties. *J. Immunol.* **161**, 1083–6.

41. Gewirtz, A. T., Simon, P. O., Jr., Schmitt, C. K., Taylor, L. J., Hagedorn, C. H., O'Brien, A. D., Neish, A. S. and Madara, J. L. (2001). *Salmonella typhimurium* translocates flagellin across intestinal epithelia, inducing a proinflammatory response. *J. Clin. Invest.* **107**, 99–109.

42. Lyons, S., Wang, L., Casanova, J. E., Sitaraman, S. V., Merlin, D. and Gewirtz, A. T. (2004). *Salmonella typhimurium* transcytoses flagellin via an SPI2-mediated vesicular transport pathway. *J. Cell Sci.* **117**, 5771–80.

43. Ramos, H. C., Rumbo, M. and Sirard, J. C. (2004). Bacterial flagellins: mediators of pathogenicity and host immune responses in mucosa. *Trends Microbiol.* **12**, 509–17.

44. Rimoldi, M., Chieppa, M., Larghi, P., Vulcano, M., Allavena, P. and Rescigno, M. (2005). Monocyte-derived dendritic cells activated by bacteria or by bacteria-stimulated epithelial cells are functionally different. *Blood* **106**, 2818–26.

45. Rakoff-Nahoum, S., Paglino, J., Eslami-Varzaneh, F., Edberg, S. and Medzhitov, R. (2004). Recognition of commensal microflora by Toll-like receptors is required for intestinal homeostasis. *Cell* **118**, 229–41.

46. Kobayashi, K., Hernandez, L. D., Galan, J. E., Janeway, C. A., Jr., Medzhitov, R. and Flavell, R. A. (2002). IRAK-M is a negative regulator of Toll-like receptor signaling. *Cell* **110**, 191–202.

47. Janssens, S., Burns, K., Tschopp, J. and Beyaert, R. (2002). Regulation of interleukin-1- and lipopolysaccharide-induced NF-kappaB activation by alternative splicing of MyD88. *Curr. Biol.* **12**, 467–71.

48. Kelly, D., Campbell, J. I., King, T. P., Grant, G., Jansson, E. A., Coutts, A. G., Pettersson, S. and Conway, S. (2004). Commensal anaerobic gut bacteria attenuate inflammation by regulating nuclear-cytoplasmic shuttling of PPAR-gamma and RelA. *Nat. Immunol.* **5**, 104–12.

49. Neish, A. S., Gewirtz, A. T., Zeng, H., Young, A. N., Hobert, M. E., Karmali, V., Rao, A. S. and Madara, J. L. (2000). Prokaryotic regulation of epithelial responses by inhibition of IkappaB-alpha ubiquitination. *Science* **289**, 1560−3.

50. Akbari, O., DeKruyff, R. H. and Umetsu, D. T. (2001). Pulmonary dendritic cells producing IL-10 mediate tolerance induced by respiratory exposure to antigen. *Nat. Immunol.* **2**, 725−31.

51. Alpan, O., Rudomen, G. and Matzinger, P. (2001). The role of dendritic cells, B cells, and M cells in gut-oriented immune responses. *J. Immunol.* **166**, 4843−52.

52. Iwasaki, A. and Kelsall, B. L. (1999). Freshly isolated Peyer's patch, but not spleen, dendritic cells produce interleukin 10 and induce the differentiation of T helper type 2 cells. *J. Exp. Med.* **190**, 229−39.

53. Williamson, E., Bilsborough, J. M. and Viney, J. L. (2002). Regulation of mucosal dendritic cell function by receptor activator of NF-kappa B (RANK)/RANK ligand interactions: impact on tolerance induction. *J. Immunol.* **169**, 3606−12.

54. Iwasaki, A. and Kelsall, B. L. (2001). Unique functions of cd11b[(+)], cd8alpha[(+)], and double-negative Peyer's patch dendritic cells. *J. Immunol.* **166**, 4884−90.

55. Sato, A., Hashiguchi, M., Toda, E., Iwasaki, A., Hachimura, S. and Kaminogawa, S. (2003). CD11b[+] Peyer's patch dendritic cells secrete IL-6 and induce IgA secretion from naive B cells. *J. Immunol.* **171**, 3684−90.

56. Stagg, A. J., Kamm, M. A. and Knight, S. C. (2002). Intestinal dendritic cells increase T cell expression of alpha4beta7 integrin. *Eur. J. Immunol.* **32**, 1445−54.

57. Mora, J. R., Bono, M. R., Manjunath, N., Weninger, W., Cavanagh, L. L., Rosemblatt, M. and Von Andrian, U. H. (2003). Selective imprinting of gut-homing T cells by Peyer's patch dendritic cells. *Nature* **424**, 88−93.

58. Johansson-Lindbom, B., Svensson, M., Wurbel, M. A., Malissen, B., Marquez, G. and Agace, W. (2003). Selective generation of gut tropic T cells in gut-associated lymphoid tissue (GALT): requirement for GALT dendritic cells and adjuvant. *J. Exp. Med.* **198**, 963−9.

59. Mora, J. R., Cheng, G., Picarella, D., Briskin, M., Buchanan, N. and von Andrian, U. H. (2005). Reciprocal and dynamic control of CD8 T cell homing by dendritic cells from skin- and gut-associated lymphoid tissues. *J. Exp. Med.* **201**, 303−16.

60. Yrlid, U. and Macpherson, G. (2003). Phenotype and function of rat dendritic cell subsets. *Apmis* **111**, 756−65.

61. Turnbull, E. L., Yrlid, U., Jenkins, C. D. and Macpherson, G. G. (2005). Intestinal dendritic cell subsets: differential effects of systemic TLR4 stimulation on migratory fate and activation in vivo. *J. Immunol.* **174**, 1374−84.

62. Smits, H. H., van Beelen, A. J., Hessle, C., Westland, R., de Jong, E., Soeteman, E., Wold, A., Wierenga, E. A. and Kapsenberg, M. L. (2004). Commensal Gram-negative bacteria prime human dendritic cells for enhanced IL-23 and IL-27 expression and enhanced Th1 development. *Eur. J. Immunol.* **34**, 1371−80.

63. Smits, H. H., Engering, A., van der Kleij, D., de Jong, E. C., Schipper, K., van Capel, T. M., Zaat, B. A., Yazdanbakhsh, M., Wierenga, E. A., van Kooyk, Y. and Kapsenberg, M. L. (2005). Selective probiotic bacteria induce IL-10-producing regulatory T cells in vitro by modulating dendritic cell function through dendritic cell-specific intercellular adhesion molecule 3-grabbing nonintegrin. *J. Allergy Clin. Immunol.* **115**, 1260−7.

64. Svensson, M., Maroof, A., Ato, M. and Kaye, P. M. (2004). Stromal cells direct local differentiation of regulatory dendritic cells. *Immunity* **21**, 805−16.

65. Zhang, M., Tang, H., Guo, Z., An, H., Zhu, X., Song, W., Guo, J., Huang, X., Chen, T., Wang, J. and Cao, X. (2004). Splenic stroma drives mature dendritic cells to differentiate into regulatory dendritic cells. *Nat. Immunol.* **5**, 1124−33.

66. Soumelis, V. and Liu, Y. J. (2004). Human thymic stromal lymphopoietin: a novel epithelial cell-derived cytokine and a potential key player in the induction of allergic inflammation. *Springer Semin. Immunopathol.* **25**, 325−33.

67. Soumelis, V., Reche, P. A., Kanzler, H., Yuan, W., Edward, G., Homey, B., Gilliet, M., Ho, S., Antonenko, S., Lauerma, A. *et al.* (2002). Human epithelial cells trigger dendritic cell mediated allergic inflammation by producing TSLP. *Nat. Immunol.* **3**, 673−80.

68. Rimoldi, M., Chieppa, M., Salucci, V., Avogadri, F., Sonzogni, A., Sampietro, G. M., Nespoli, A., Viale, G., Allavena, P. and Rescigno, M. (2005). Intestinal immune homeostasis is regulated by the crosstalk between epithelial cells and dendritic cells. *Nat. Immunol.* **6**, 507−14.

69. Kosiewicz, M. M., Nast, C. C., Krishnan, A., Rivera-Nieves, J., Moskaluk, C. A., Matsumoto, S., Kozaiwa, K. and Cominelli, F. (2001). Th1-type responses mediate spontaneous ileitis in a novel murine model of Crohn's disease. *J. Clin. Invest.* **107**, 695−702.

70. Mastroeni, P. and Menager, N. (2003). Development of acquired immunity to *Salmonella*. *J. Med. Microbiol.* **52**, 453−9.

71. Hess, J., Ladel, C., Miko, D. and Kaufmann, S. H. (1996). *Salmonella typhimurium* aroA− infection in gene-targeted immunodeficient mice: major role of CD4$^+$ TCR-alpha beta cells and IFN-gamma in bacterial clearance independent of intracellular location. *J. Immunol.* **156**, 3321−6.

72. George, A. (1996). Generation of gamma interferon responses in murine Peyer's patches following oral immunization. *Infect. Immun.* **64**, 4606−11.

73. Liesenfeld, O., Kosek, J. C. and Suzuki, Y. (1997). Gamma interferon induces Fas-dependent apoptosis of Peyer's patch T cells in mice following peroral infection with *Toxoplasma gondii*. *Infect. Immun.* **65**, 4682−9.

74. Vossenkamper, A., Struck, D., Alvarado-Esquivel, C., Went, T., Takeda, K., Akira, S., Pfeffer, K., Alber, G., Lochner, M., Forster, I. and Liesenfeld, O. (2004). Both IL-12 and IL-18 contribute to small intestinal Th1-type immunopathology following oral infection with *Toxoplasma gondii*, but IL-12 is dominant over IL-18 in parasite control. *Eur. J. Immunol.* **34**, 3197−207.

CHAPTER 5

Role of dendritic cells in the innate response to bacteria

Natalya V. Serbina and Eric G. Pamer
Sloan Kettering Institute

5.1 INTRODUCTION

Innate immunity is an ancient and highly conserved system that provides the first line of defense upon encounter with pathogenic organisms. Activation of innate immune responses is a complex process involving multiple components and distinct steps. The cellular components of innate immunity include neutrophils, monocytes, macrophages and dendritic cells (DCs). These cells are capable of direct microbicidal activity that partially depends on inducible nitric synthase (iNOS) and NADPH oxidase complex that catalyze production of toxic anti-microbial compounds[1,2]. Additionally, they secrete a vast array of pro-inflammatory mediators such as cytokines and chemokines and can recruit and activate other inflammatory cells, thus amplifying the immune cascade. Apart from their role in restricting microbial growth, innate immune responses also provide the inflammatory context in which adaptive T- and B-cell immune responses develop.

Dendritic cells are derived from hematopoietic progenitor cells in the bone marrow and are found in the peripheral circulation as well as in the lymphoid and non-lymphoid tissues. Dendritic cells can be subdivided into several subsets based on the expression of the cell surface markers and different subsets have been ascribed distinct functions during the immune response[3]. Since their discovery, dendritic cells have been studied extensively with regard to their role as antigen-presenting cells[4]. However, it is becoming increasingly clear that dendritic cells also play an important role during the innate immune responses to microbial pathogens.

Dendritic cells are found in peripheral tissues and are strategically positioned at the sites where pathogen encounters are most frequent, such as gut, lung and skin. Dendritic cells at these different sites are morphologically and functionally heterogeneous, potentially optimizing interactions with pathogens in the context of their local microenvironment. For example, microbes invading through breaks in the skin encounter epidermal Langerhans cells and dermal dendritic cells that are highly phagocytic and can migrate to lymph nodes under inflammatory conditions[5]. In the intestinal mucosae, lamina propria dendritic cells express tight junction proteins and can open junctions between epithelial cells, thereby enabling sampling of antigens in the intestinal lumen[6]. Pathogens entering via the respiratory route are detected by dendritic cells residing within the epithelium and interstitium of the lung. Microbial contact generally induces activation of lung dendritic cells and their migration to lung-draining lymph nodes[7].

Dendritic cell uptake of microbes at the epithelial and mucosal sites and their subsequent migration to lymphoid organs may contribute to dissemination of microbes throughout the body. During intestinal *Salmonella typhimurium* infection, spread of bacteria to extraintestinal sites requires CD18 expressing phagocytes[8]. These CD18-positive cells transport *Salmonella* from intestinal epithelium through the bloodstream to the spleen. Lung dendritic cells can be infected with *Mycobacterium tuberculosis*[9] and phagocytose bacteria in vivo, albeit at significantly lower levels than macrophages[10]. Migration of infected dendritic cells to lung-draining lymph nodes has been suggested to trigger development of T cell immunity as well as contribute to bacterial dissemination.

Dendritic cells express a wide range of TLRs and can recognize and respond to distinct classes of pathogens[11]. Additionally, dendritic cells also express Nod proteins and may be capable of detecting microbes replicating in cytosol[12,13]. Infection with live bacteria or stimulation with bacterial components induces dendritic cells to secrete cytokines and chemokines, thus directly contributing to the inflammatory milieu. Furthermore, dendritic cells can orchestrate immune responses by directly activating other immune cells such as NK cells, T and B cells. Additionally, activated dendritic cells can secrete nitric oxide and thus may have a direct bactericidal activity. It was demonstrated that microbial stimulation of dendritic cells induces very early changes in their gene expression profile suggesting that they might play a role during the early onset of inflammatory responses[12,13].

Thus, the role of dendritic cells in the innate immune responses is multifaceted and involves distinct effector functions. Microbial recognition, in vivo bacterial uptake by dendritic cells and activation of other innate immune cells are addressed elsewhere in this book. In this chapter, we will summarize recent advances in our understanding of the role of dendritic cells in microbial killing, focusing in particular on the murine model of *Listeria monocytogenes* infection.

5.2 PATHOGENESIS OF *LISTERIA MONOCYTOGENES*

Listeria monocytogenes is a pathogenic Gram-positive facultative intra-cellular bacterium and is the cause of listeriosis. *Listeria* can infect humans and risk factors for infection include pregnancy and the immunocom-promised state that follows chemotherapy or organ transplantation[14]. Murine infection with *L. monocytogenes* has been used extensively as a model for studying host immune responses to intracellular infection and has been instrumental in delineating various components of the innate and adaptive arms of anti-microbial immunity.

The natural route of *Listeria* infection is via the gastrointestinal tract when food contaminated with bacteria is ingested. *Listeria* can infect a variety of phagocytic and non-phagocytic cells. Invasion of cells of the intestinal epithelium requires interaction of bacterially expressed internalin A (IntA) with epithelial E-cadherin molecules[15]. Another internalin molecule, Int B, enables bacteria to invade hepatocytes in the liver[16]. Several receptors have been shown to interact with IntB such as glycosaminoglycans, hepatocyte growth factor receptor and complement receptors[17–20]. Uptake of *Listeria* by phagocytic cells can involve macrophage scavenger receptors [21] and complement receptors [22] on the surface of the host cells.

Inside the infected cells, bacteria are initially found in vacuoles but rapidly escape into the cytoplasm by secreting membranolytic protein, listeriolysin O (LLO)[23]. In some cell types cytoplasmic invasion also involves the action of bacterially encoded phospholipases[24]. Expression of LLO is essential for bacterial virulence and *L. monocytogenes* strains deficient in this molecule are highly attenuated. Invasion of the cytosol by bacteria triggers innate inflammatory responses[24,25] and appears to be required for induction of protective immunity[26]. In the cytoplasm, bacteria express the actin-assembly-inducing protein ActA, enabling bacterial movement through the cytoplasm and into neighboring cells[27,28]. ActA is a bacterial virulence factor and ActA-deficient *L. monocytogenes* strain are markedly attenuated. Interestingly, while both ActA-deficient and LLO-deficient strains are attenuated in mice, infection with ActA-deficient bacteria induces protective long-term immunity[29] and, as noted before, infection with LLO-deficient strain does not[26]. Thus, protective immunity can be dissociated from virulence. Cytosol invasion by bacteria establishes the inflammatory environment that optimizes the development of adaptive T cell responses, potentially by activating dendritic cells.

5.3 KEY MEDIATORS OF INNATE IMMUNE RESPONSES TO *L. MONOCYTOGENES*

During *L. monocytogenes* infection, both innate and adaptive immunity have been shown to contribute to bacterial clearance. While complete in vivo clearance of *L. monocytogenes* is T cell dependent, early control of bacterial replication depends on induction of innate immunity[30,31]. Inbred strains of mice differ with respect to their innate resistance to infection[32,33]. Genetic analysis indicates that resistance is a complex genetic trait; several loci on chromosomes 1[34], 2[35,36], 5 and 13[34] have been linked to host resistance against *Listeria* infection. Following infection, innate immune responses are triggered rapidly[37]. The importance of innate immunity is indicated by the demonstration that mice lacking both CD4 and CD8 T cells are, for several weeks, able to control infection, although they cannot clear it in the long term[31,38]. Innate immune responses to *Listeria* have been extensively studied and various immune components have been delineated. Studies with gene-deficient mice demonstrated that both TNF-α and IFN-γ are crucial for immune protection during infection[39,40]. Mice lacking these cytokines or their respective receptors succumb to infection within a few days, indicating failure of innate immunity[41-43]. Although the pivotal role of these cytokines during anti-listerial innate immune response is established, their cellular sources and their down-stream effects are not well understood. It is thought that early production of IFN-γ by natural killer cells activates the bactericidal effector functions of macrophages, thus leading to bacterial killing. However, mice that lack NK cells are resistant to infection and have normal innate immunity which suggests that there are NK-independent sources of IFN-γ[44].

Infected macrophages have long been considered to be a major source of TNF-α. Although the role of TNF-α in the activation and regulation of innate inflammatory responses is well recognized, its precise mode of action in defense against *Listeria* infection remains unresolved. Mice that express a non-sheddable p55TNFR show innate immune hyperactivity and are more resistant to *L. monocytogenes*; however, failure to downregulate TNF receptor also leads to inflammatory pathology[45]. Additionally, both IL-12 and IL-18 contribute to protection against *L. monocytogenes*, possibly through induction of other cytokines[46,47]. In contrast, mice lacking

Type I interferon receptor or the IRF3 transcription factor have enhanced resistance to *L. monocytogenes* infection[48-50]. In theses studies, signaling through Type I interferon receptors was suggested to enhance susceptibility of T cells to apoptosis[48,49], down-modulate IL-12 production and decrease numbers of TNF producing cells[50]. Thus, induction of innate immune responses during intracellular bacterial infection is a complex process involving multiple regulatory steps.

5.4 ROLE OF MYELOID CELLS

Myeloid cells are comprised of distinct subpopulations, all of which have innate immune functions. Granulocytes, monocytes and dendritic cell precursors are derived from granulocyte−monocyte progenitor cells in the bone marrow. During immune responses to bacterial infections, neutrophils and monocytes play a crucial role. Because of their common lineage, these cells share number of cell surface receptors and adhesion molecules such as CD11b, CD11c, Gr-1 and Mac-3 and have overlapping effector functions. From an immunologist perspective, the common expression of multiple surface markers makes it difficult to distinguish the distinct myeloid subpopulations during the innate immune responses.

During innate immune responses to *L. monocytogenes*, cells of myeloid lineage play a key role[51]. The contributions of granulocytes and monocytes have been characterized in antibody depletion studies. In vivo administration of RB6-8C5 antibody specific for Gr-1 renders mice highly susceptible to *L. monocytogenes* infection[52-54]. The deleterious effect of RB6-8C5 antibody administration is particularly pronounced when antibody is given during the first two days of infection[53,54]. RB6-8C5 antibody reacts with the common epitope on Ly6G and Ly6C antigens expressed on monocytes and dendritic cells as well as on neutrophils. Thus, while neutrophils are undoubtedly indispensable for the early control of bacterial replication, the potential contribution of other Gr-1-expressing myeloid cells cannot be discounted.

Administration of a blocking monoclonal antibody specific for type 3 complement receptor (CD11b, Mac-1) impairs immunity of mice to *L. monocytogenes* and leads to increased mortality in response to a sublethal dose of bacteria[55]. Similar to Gr-1 treatment, anti-CD11b treatment interferes with immune responses only when given early during infection. CD11b blocking by antibody leads to a failure to focus monocytic cells at sites of infection and results in unrestricted bacterial multiplication in organs, suggesting that CD11b-expressing cells are essential for bacterial containment and killing during early infection. The CD11b-bearing myelo-monocytic population is comprised of neutrophils, macrophages and dendritic cells; while all of the cells can contribute to innate immune responses, their specific contributions to bacterial clearance remain undefined.

The functions of macrophages and granulocytes during infection have been extensively studied. It is thought that these cells exert bactericidal activity by producing reactive nitrogen and oxygen intermediates[56] as well as through the action of phagolysosomal enzymes[57]. Efficient containment of *L. monocytogenes* in the vacuoles of activated macrophages and killing of bacteria in vitro depends on production of ROI and RNI[51,58]. Mice lacking iNOS[58,59] and components of the NADPH oxidase complex have increased susceptibility to *L. monocytogenes*[60,61] suggesting that these mechanisms contribute to bacterial clearance in vivo.

5.5 TipDCs IN THE INNATE IMMUNE RESPONSES TO *L. MONOCYTOGENES*

The direct demonstration that dendritic cells play a role during in vivo innate immune responses to *Listeria* came from recent studies that examined course of infection in mice lacking chemokine receptor CCR2. CCR2 is expressed on monocytes and dendritic cells and is implicated in the in vivo migration of these cells under inflammatory conditions[62-72]. Immunity to *L. monocytogenes* is profoundly diminished in CCR2-deficient mice; knockout mice succumb to infection within 4 days indicating failure of innate immunity[63]. Since CCR2 is expressed by circulating monocytes, it has been suggested that the inability to recruit macrophages to the sites of infection leads to impaired immune responses in CCR2-deficient mice. Analysis of innate immune responses in CCR2-deficient mice revealed markedly diminished levels of TNF and iNOS. It has been postulated that *L. monocytogenes* infected macrophages are the major source of these molecules during infection. Characterization of myeloid populations in the spleens of infected mice revealed that production of TNF and iNOS was largely attributable to a novel monocyte-derived dendritic cell population (TNF/iNOS producing dendritic cells, TipDCs) that is recruited to sites of bacterial replication in a CCR2-dependent fashion. Phenotypically, TipDCs express of CD11b, CD11c and high levels of intracellular Mac-3, markers that are also expressed by some conventional dendritic cells. TipDCs also express high levels of MHC Class II and co-stimulatory molecules and expression of these molecules is further induced during the course of infection. In contrast to conventional dendritic cells, TipDcs do not express CD4, CD8 or CD205. Interestingly, TipDCs also express Gr-1, a marker expressed by plasmacytoid dendritic cell subset. However, TipDCs appear to be distinct from plasmacytoid population since they do not express B220 and CCR2 knockout mice have normal levels of Type I IFNs.

To confirm that TNF- and iNOS-producing cells represent a population of dendritic cells and not macrophages, their ability to prime T cell responses was examined. TipDCs induced proliferation of naive allogenic T cells in in vitro cultures, albeit at lower levels than conventional DCs. Interestingly, CCR2 deficient mice develop enhanced CD8 T cell responses (see below) following infection with attenuated strains of *L. monocytogenes*. Thus, TipDCs do not appear to contribute to T cell priming in vivo and have a distinct role during infection.

The essential role of TipDCs in the anti-listerial immune response is attributed to their ability to produce TNF and nitric oxide during infection. Intracellular cytokine analysis revealed that TipDCs are capable of producing copious amounts of TNF in response to stimulation with *Listeria* and secrete much more TNF than neutrophils and monocytes. In the infected spleen, TipDCs appear to be the only major source of iNOS; consequently, nitric oxide production is drastically reduced in CCR2-deficient spleens. It remains to be determined whether the ability of TipDCs to express high levels of iNOS and TNF is due to the unique expression of receptors or signaling molecules in this subset. Interestingly, although TipDCs are found in the close proximity to cells harboring bacteria in vivo, they do not appear to harbor significant numbers of bacteria in vivo. It is possible that bacteria are rapidly killed and degraded by TipDCs upon phagocytosis in which case very few cells would appear directly infected. Alternatively, TipDCs may impact bacterial killing by orchestrating innate immune responses via production of TNF and nitric oxide without being directly invaded. NO is a water- and lipid-soluble gas and it might exert anti-microbial effects by diffusing into infected cells in the vicinity of Tip-DCs. Additionally, exogenous NO can induce iNOS expression in infected cells[73]. In the absence of activation, infection of macrophages with live *L. monocytogenes* leads to the rapid destruction of host cells Therefore, it may be beneficial to recruit cells capable of sensing infection and mediating microbicidal effector functions without being directly infected.

In addition to direct bactericidal activity, nitric oxide production by TipDCs might have other regulatory functions in the setting of innate immunity. For example, nitric oxide production has inhibitory effect on

T cell proliferation[74,75]. Interestingly, CD8 T cell responses in the spleens of CCR2-deficient mice are larger than those observed in the wild type spleens; thus, TipDCs might regulate the magnitude of developing T cell responses.

The CCR2-expressing monocytes circulating in the blood of naive mice and recruited to the sites of inflammation via CCR2 have recently been described[69]. After migration to the inflamed tissues, these cells have a capacity to differentiate into dendritic cells. Similar to TipDCs, the inflammatory monocytes express Gr-1 and CD11b and do not express CD4 and CD8 molecules. Thus, it is conceivable that these circulating cells represent precursors of TipDCs.

TNF secretion by TipDCs requires MyD88-mediated signaling but is intact in the absence of TLR2[25]. It is possible that the combination of multiple distinct TLRs and/or the levels of receptor expression determine the capacity of TipDCs to produce these inflammatory mediators.

It remains to be determined whether TipDCs play a role during immune responses to other bacterial pathogens. In vitro infection of dendritic cells with *M. tuberculosis* and *S. typhimurium* induces iNOS expression and secretion of nitric oxide[76–79]. Recently, recruitment of CCR2+CD11cint dendritic cells to the lungs of mice infected with *M. tuberculosis* has been reported[70]. TNF and iNOS production by lung CD11c intermediate cells was not examined in this study and it remains to be determined whether they represent a TipDC population. Another study reported the presence of monocyte-derived dendritic cells capable of bactericidal activity against intracellular *M. tuberculosis* in bronchoalveolar lavage samples[80]. At this point, further studies are needed to determine whether recruitment of TipDCs represents a common mechanism for defense against intracellular bacterial infections.

5.6 DENDRITIC CELL INTERACTION WITH *L. MONOCYTOGENES*

Much of what is known regarding dendritic cell interaction with pathogens has been learned from in vitro studies. Human and murine dendritic cells can be derived by culturing blood or bone marrow monocytes with GM-CSF in the presence or absence of IL-4. The resulting cell populations express high levels of MHC Class II and co-stimulatory molecules. GM-CSF differentiated cells have been very instrumental in examining dendritic cell interactions with T cells and impact of bacterial infection on this cell type.

Human and murine dendritic cells can be infected with *L. monocytogenes* in vitro[81–83]. Bacterial internalization occurs independent of internalin A and B and involves glycosylated receptors on the surface of dendritic cells[81]. Whether *Listeria* escape vacuoles and replicate in the cytoplasm of infected dendritic cells is controversial with reports demonstrating bacterial escape into cytoplasm of murine dendritic cells[82,83], while another study reported that internalized *Listeria* are contained predominantly in phagosomes[81]. Interestingly, although bacteria induce killing of murine dendritic cells, a small proportion of infected cells remains infected for a prolonged period of time[83]. In other infectious disease settings, the ability of dendritic cells to harbor live bacteria for a prolonged period of time has been suggested to contribute to bacterial dissemination and T cell priming[80,84–86]. Invasion of immature dendritic cells by *L. monocytogenes* induces their maturation and enhanced expression of MHC Class II and co-stimulatory molecules[81,87]; in one study this effect appeared to be largely due to listerial lipoteichoic acid[81]. In addition to inducing dendritic cell maturation, infection also stimulates in vitro cytokine production by DCs[88,89].

Activation of the innate immune cells and subsequent cytokine secretion occurs following triggering of the pathogen-recognition receptors, Toll-like receptors (TLRs, see Chapter 2). Various members of the TLR family recognize evolutionary conserved, pathogen-derived components such as lipopolysaccharide (LPS), peptidoglycan (PGN), lipoteichoic acid (LTA), CpG DNA, bacterial flagellin and viral double-stranded RNA. TLR family members signal via the MyD88 adaptor molecule[90]. Signaling through TLR-MyD88 pathway is critical for initiation of antimicrobial immune responses and Myd88-deficient mice are highly susceptible to a number of bacterial pathogens. Immune responses to *L. monocytogenes* have been examined in mice lacking specific innate immune receptors as well as MyD88. Genetic deletion of Myd88 in mice leads to dramatically diminished resistance to *L. monocytogenes*[91,92]. Moreover, MyD88-deficient mice display greater susceptibility than mice lacking IFN-γ or both IL-12 and IFN-γ[91] IL-1 and IL-18-signaling is MyD88-dependent[93]. Impaired immunity in the absence of MyD88 is attributed to loss of TLR-mediated signals since mice deficient in Caspase-1 (and therefore secreted forms of IL-1 and IL-18) are only marginally susceptible to *Listeria*[92,94].

Listeria can be recognized through distinct TLRs such as TLR2, receptor for petidoglycan, lipoproteins and lipoteichoic acid, TLR5, receptor for flagellin and TLR9, receptor for unmethylated CpG DNA. Infection with *L. monocytogenes* induces IL-12 and IL-18 secretion by human DCs[109] while stimulation with listerial LTA induced secretion of IL-18 but only minimal IL-12, suggesting that live bacteria are recognized by multiple TLRs.

Recognition of *L. monocytogenes* through TLR2 and TLR5 has been documented[91,95]. However, conflicting results have been reported regarding the role of TLR2 in innate immune responses to *L. monocytogenes* infection[25,91,92] and in vivo data on TLR5 is still lacking. Recently, administration of immunostimulatory CpG oligodeoxynucleotides was shown to render newborn mice resistant to *L. monocytogenes* infection. In CpG-treated mice, dendritic cells, macrophages and B cells responded to stimulation by secreting IFN-γ, TNF-α and IL-12[96]. It remains to be established whether TLR9-mediated recognition of *Listeria* contributes to the immune responses in vivo.

Dendritic cell maturation and production of IL-12, IL-10, IL-6 and TNF-α is efficiently triggered by cytosol-invasive and not LLO-deficient bacteria indicating that dendritic cells are capable of sensing the presence of *Listeria* in the cytoplasm[87]. Recently, microbial recognition by cytoplasmic proteins Nod1 and Nod2 has been reported. Nod1 and Nod2 belong to NBS-LRR family of mammalian proteins and recognize bacterial petidoglycan motifs associated with Gram-positive and Gram-negative cell wall. Nod2 recognizes muramyl dipeptide (MDP), a conserved structure present in the cell wall of Gram-positive and Gram-negative bacteria[97–99]. Nod 1 detects a unique diaminopimelate-containing *N*-acetylglucosamine-*N*-acetylmuramic acid tripeptide motif present in the Gram-negative bacterial peptidoglycan[100]. Dendritic cells express Nod1 and Nod2 and release cytokines in response to stimulation with muropeptides[12,13]. Nod2-deficient mice have increased susceptibility to *L. monocytogenes* administered via the intragastric route[101], suggesting that *Listeria*-derived cell wall components are recognized by Nod2 in vivo. Additionally, the cell wall of *L. monocytogenes* contains mesoDAP and thus can potentially be recognized by Nod1. Although, extracts of *L. monocytogenes* do not induce Nod1-mediated signaling in vitro[98], the potential role of Nod1 recognition during in vivo infection cannot be ruled out. Both Nod1 and Nod2 associate with the serine/threonine kinase RIP2 and their signaling leads to NF-κB activation. The role of Rip2 in defense against *L. monocytogenes* has been examined. IL-6 production by *L. monocytogenes* infected macrophages is diminished in the absence of Rip2 and Rip2-deficient mice are more susceptible to in vivo infection[102,103]. Rip2 is positioned downstream of TLR and Nod proteins in the signaling cascade. However, MyD88-deficient mice are significantly more susceptible to *Listeria* infection than Rip2-deficient mice, suggesting that Rip2 plays a more limited role in the TLR-mediated immune activation. In vitro infection of cells with *L. monocytogenes* induces NF-κB activation[104], production of type I IFNs[24], and MCP-1[25] that is dependent on bacterial localization to cytoplasm. A recent report demonstrated that secretion of type I IFNs induced by cytoplasmic bacteria requires serine-threonine kinase TNFR-associated NF-κB kinase (TANK-binding kinase 1 (TBK1))[105]. Further studies are required to address the contribution of these signaling molecules to the in vivo activation of dendritic cells and macrophages during innate immune responses to *L. monocytogenes*.

5.7 IN VIVO FUNCTION OF DENDRITIC CELLS

In lymphoid organs, dendritic cells can be divided into CD8-positive and CD8-negative populations; the CD8-negative population can be further subdivided based on their expression of CD4 and CD205[3]. CD8-positive dendritic cells express CD11c, CD205 and high levels of MHC Class II and do not express CD11b. CD8-negative dendritic cell subsets express varied levels of CD11b and Gr-1. Due to the fact that many cell surface markers are shared between dendritic cell subsets and other cells of myeloid lineage, the contribution of distinct subpopulations to immune responses and bacterial clearance in vivo is difficult to assess.

Dendritic cells are infected with *L. monocytogenes* in vivo after intestinal inoculation[106]. Listeria-loaded dendritic cells are first detected beneath the epithelial lining of Peyer's patches; 6 hours following infection, bacteria-containing cells are present in the draining lymph nodes suggesting that dendritic cells may be involved in bacterial dissemination. The role of dendritic cells during in vivo *L. monocytogenes* infection has been addressed in mice treated with Flt-3 ligand, a hemapoietic growth factor that drastically increases the numbers of DCs in organs. Treatment of neonatal mice with Flt-3 ligand increases the numbers of CD11c⁺MHC Class II^high and CD11c⁺B220⁺ dendritic cells in the spleen, liver, skin and peritoneum and enhances resistance to *L. monocytogenes* in IL-12-dependent manner suggesting that dendritic cells play an active part during innate immune responses[107]. However, different results were reported by another study in which Flt-3 ligand was administered to adult mice and this treatment led to impaired protective immunity despite enhanced T cell priming[110]. Although these studies provide the insight into our understanding of potential functions of dendritic cells, they are based on the artificial enhancement of dendritic cell numbers and may not reflect the immune responses following infection in vivo.

Recently, the in vivo role of dendritic cells in the priming of adaptive immune responses has been examined using transient in vivo depletion of CD11c-expressing cells[108]. Utilizing diphtheria toxin-based system,

Jung *et al.* demonstrated that short-term depletion of CD11c-expressing dendritic cells abrogates in vivo CD8 T cell priming during *L. monocytogenes* infection. It remains to be determined whether T cell priming by dendritic cells during infection is carried out by a small fraction of cells that are directly infected or whether it involves antigen cross-presentation. This study reinforced the idea that potent in vitro antigen-presenting capacity of dendritic cells can be directly translated to their in vivo role. The impact of CD11c$^+$ cell depletion on the development of innate immune responses to *L. monocytogenes* remains unclear.

5.8 CONCLUSIONS

In the past two decades, research in the dendritic cell field has been centered on the antigen presenting capacity of these cells. However, it is becoming increasingly clear that dendritic cells contribute to many aspects of innate inflammatory responses. Understanding the molecular and cellular mechanisms of dendritic cell functions during innate immune responses is important for better understanding of the pathogenesis of infectious diseases as well as immune-related disorders and will lead to the development of novel immunologic therapies.

REFERENCES

1. Taylor, P. R. *et al.* (2005). Macrophage receptors and immune recognition. *Annu. Rev. Immunol.* **23**, 901−44.
2. Segal, A. W. (2005). How neutrophils kill microbes. *Annu. Rev. Immunol.* **23**, 197−223.
3. Shortman, K. and Y. J. Liu (2002). Mouse and human dendritic cell subtypes. *Nat. Rev. Immunol.* **2**(3), 151−61.
4. Steinman, R. M. (1991). The dendritic cell system and its role in immunogenicity. *Annu. Rev. Immunol.* **9**, 271−96.
5. Carbone, F. R., G. T. Belz, and W. R. Heath (2004). Transfer of antigen between migrating and lymph node-resident DCs in peripheral T-cell tolerance and immunity. *Trends Immunol.* **25**(12), 655−8.
6. Rescigno, M. *et al.* (2001). Dendritic cells express tight junction proteins and penetrate gut epithelial monolayers to sample bacteria. *Nat. Immunol.* **2**(4), 361−7.
7. Lambrecht, B. N., J. B. Prins, and H. C. Hoogsteden (2001). Lung dendritic cells and host immunity to infection. *Eur. Respir. J.* **18**(4), 692−704.

8. Vazquez-Torres, A. *et al.* (1999). Extraintestinal dissemination of *Salmonella* by CD18-expressing phagocytes. *Nature* **401**(6755), 804–8.

9. Gonzalez-Juarrero, M. and I. M. Orme (2001). Characterization of murine lung dendritic cells infected with *Mycobacterium tuberculosis. Infect. Immun.* **69**(2), 1127–33.

10. Pedroza-Gonzalez, A. *et al.* (2004). *In situ* analysis of lung antigen-presenting cells during murine pulmonary infection with virulent *Mycobacterium tuberculosis. Int. J. Exp. Pathol.* **85**(3), 135–45.

11. Iwasaki, A. and R. Medzhitov (2004). Toll-like receptor control of the adaptive immune responses. *Nat. Immunol.* **5**(10), 987–95.

12. Fritz, J. H. *et al.* (2005). Synergistic stimulation of human monocytes and dendritic cells by Toll-like receptor 4 and NOD1- and NOD2-activating agonists. *Eur. J. Immunol.* **35**(8), 2459–70.

13. Gutierrez, O. *et al.* (2002). Induction of Nod2 in myelomonocytic and intestinal epithelial cells via nuclear factor-kappa B activation. *J. Biol. Chem.* **277**(44), 41701–5.

14. Gellin, B. G. and C. V. Broome (1989). *Listeriosis. JAMA* **261**(9), 1313–20.

15. Gaillard, J. L. *et al.* (1991). Entry of *L. monocytogenes* into cells is mediated by internalin, a repeat protein reminiscent of surface antigens from gram-positive cocci. *Cell* **65**(7), 1127–41.

16. Dramsi, S. *et al.* (1995). Entry of *Listeria monocytogenes* into hepatocytes requires expression of inIB, a surface protein of the internalin multigene family. *Mol. Microbiol.* **16**(2), 251–61.

17. Braun, L., B. Ghebrehiwet, and P. Cossart (2000). gC1q-R/p32, a C1q-binding protein, is a receptor for the InlB invasion protein of *Listeria monocytogenes. EMBO J.* **19**(7), 1458–66.

18. Shen, Y. *et al.* (2000). InIB-dependent internalization of *Listeria* is mediated by the Met receptor tyrosine kinase. *Cell* **103**(3), 501–10.

19. Marino, M. *et al.* (2002). GW domains of the *Listeria monocytogenes* invasion protein InlB are SH3-like and mediate binding to host ligands. *EMBO J.* **21**(21), 5623–34.

20. Jonquieres, R., J. Pizarro-Cerda, and P. Cossart (2001). Synergy between the N- and C-terminal domains of InlB for efficient invasion of non-phagocytic cells by *Listeria monocytogenes. Mol. Microbiol.* **42**(4), 955–65.

21. Dunne, D. W. *et al.* (1994). The type I macrophage scavenger receptor binds to Gram-positive bacteria and recognizes lipoteichoic acid. *Proc. Natl Acad. Sci. U S A* **91**(5), 1863–7.

22. Drevets, D. A. and P. A. Campbell (1991). Roles of complement and complement receptor type 3 in phagocytosis of *Listeria monocytogenes* by

inflammatory mouse peritoneal macrophages. *Infect. Immun.* **59**(8), 2645−52.

23. Bielecki, J. *et al.* (1990). *Bacillus subtilis* expressing a haemolysin gene from *Listeria monocytogenes* can grow in mammalian cells. *Nature* **345**(6271), 175−6.

24. O'Riordan, M. *et al.* (2002). Innate recognition of bacteria by a macrophage cytosolic surveillance pathway. *Proc. Natl Acad. Sci. U S A* **99**(21), 13861−6.

25. Serbina, N. V. *et al.* (2003). Sequential MyD88-independent and -dependent activation of innate immune responses to intracellular bacterial infection. *Immunity* **19**(6), 891−901.

26. Berche, P., J. L. Gaillard, and P. J. Sansonetti (1987). Intracellular growth of *Listeria monocytogenes* as a prerequisite for in vivo induction of T cell-mediated immunity. *J. Immunol.* **138**(7), 2266−71.

27. Domann, E. *et al.* (1992). A novel bacterial virulence gene in *Listeria monocytogenes* required for host cell microfilament interaction with homology to the proline-rich region of vinculin. *EMBO J.* **11**(5), 1981−90.

28. Kocks, C. *et al.* (1992). *L. monocytogenes*-induced actin assembly requires the actA gene product, a surface protein. *Cell* **68**(3), 521−31.

29. Goossens, P. L. and G. Milon (1992). Induction of protective CD8^{+} T lymphocytes by an attenuated *Listeria monocytogenes* actA mutant. *Int. Immunol.* **4**(12), 1413−18.

30. Harty, J. T. and M. J. Bevan (1996). CD8 T-cell recognition of macrophages and hepatocytes results in immunity to *Listeria monocytogenes*. *Infect. Immun.* **64**(9), 3632−40.

31. Bancroft, G. J., R. D. Schreiber, and E. R. Unanue (1991). Natural immunity: a T-cell-independent pathway of macrophage activation, defined in the scid mouse. *Immunol. Rev.* **124**, 5−24.

32. Stevenson, M. M., P. A. Kongshavn, and E. Skamene (1981). Genetic linkage of resistance to *Listeria monocytogenes* with macrophage inflammatory responses. *J. Immunol.* **127**(2), 402−7.

33. Cheers, C. and I. F. McKenzie (1978). Resistance and susceptibility of mice to bacterial infection: genetics of listeriosis. *Infect. Immun.* **19**(3), 755−62.

34. Boyartchuk, V. *et al.* (2004). The host resistance locus sst1 controls innate immunity to *Listeria monocytogenes* infection in immunodeficient mice. *J. Immunol.* **173**(8), 5112−20.

35. Gervais, F., C. Desforges, and E. Skamene (1989). The C5-sufficient A/J congenic mouse strain. Inflammatory response and resistance to *Listeria monocytogenes*. *J. Immunol.* **142**(6), 2057−60.

36. Gervais, F., M. Stevenson, and E. Skamene (1984). Genetic control of resistance to *Listeria monocytogenes*: regulation of leukocyte inflammatory responses by the Hc locus. *J. Immunol.* **132**(4), 2078–83.

37. Unanue, E. R. (1997). Studies in listeriosis show the strong symbiosis between the innate cellular system and the T-cell response. *Immunol. Rev.* **158**, 11–25.

38. Nickol, A. D. and P. F. Bonventre (1977). Anomalous high native resistance to athymic mice to bacterial pathogens. *Infect. Immun.* **18**(3), 636–45.

39. Buchmeier, N. A. and R. D. Schreiber (1985). Requirement of endogenous interferon-gamma production for resolution of *Listeria monocytogenes* infection. *Proc. Natl Acad. Sci. U S A* **82**(21), 7404–8.

40. Havell, E. A. (1989). Evidence that tumor necrosis factor has an important role in antibacterial resistance. *J. Immunol.* **143**(9), 2894–9.

41. Pfeffer, K. *et al.* (1993). Mice deficient for the 55 kd tumor necrosis factor receptor are resistant to endotoxic shock, yet succumb to *L. monocytogenes* infection. *Cell* **73**(3), 457–67.

42. Rothe, J. *et al.* (1993). Mice lacking the tumour necrosis factor receptor 1 are resistant to TNF-mediated toxicity but highly susceptible to infection by *Listeria monocytogenes*. *Nature* **364**(6440), 798–802.

43. Harty, J. T. and M. J. Bevan (1995). Specific immunity to *Listeria monocytogenes* in the absence of IFN gamma. *Immunity* **3**(1), 109–17.

44. Andersson, A. *et al.* (1998). Early IFN-gamma production and innate immunity during *Listeria monocytogenes* infection in the absence of NK cells. *J. Immunol.* **161**(10), 5600–6.

45. Xanthoulea, S. *et al.* (2004). Tumor necrosis factor (TNF) receptor shedding controls thresholds of innate immune activation that balance opposing TNF functions in infectious and inflammatory diseases. *J. Exp. Med.* **200**(3), 367–76.

46. Tripp, C. S. *et al.* (1994). Neutralization of IL-12 decreases resistance to *Listeria* in SCID and C.B-17 mice. Reversal by IFN-gamma. *J. Immunol.* **152**(4), 1883–7.

47. Neighbors, M. *et al.* (2001). A critical role for interleukin 18 in primary and memory effector responses to *Listeria monocytogenes* that extends beyond its effects on interferon gamma production. *J. Exp. Med.* **194**(3), 343–54.

48. Carrero, J. A., B. Calderon, and E. R. Unanue (2004). Type I interferon sensitizes lymphocytes to apoptosis and reduces resistance to listeria infection. *J. Exp. Med.* **200**(4), 535–40.

49. O'Connell, R. M. *et al.* (2004). Type I interferon production enhances susceptibility to *Listeria monocytogenes* infection. *J. Exp. Med.* **200**(4), 437–45.

50. Auerbuch, V. *et al.* (2004). Mice lacking the type I interferon receptor are resistant to *Listeria monocytogenes. J. Exp. Med.* **200**(4), 527–33.

51. Beckerman, K. P. *et al.* (1993). Release of nitric oxide during the T cell-independent pathway of macrophage activation. Its role in resistance to *Listeria monocytogenes. J. Immunol.* **150**(3), 888–95.

52. Conlan, J. W. and R. J. North (1994). Neutrophils are essential for early anti-*Listeria* defense in the liver, but not in the spleen or peritoneal cavity, as revealed by a granulocyte-depleting monoclonal antibody. *J. Exp. Med.* **179**(1), 259–68.

53. Rogers, H. W. and E. R. Unanue (1993). Neutrophils are involved in acute, nonspecific resistance to *Listeria monocytogenes* in mice. *Infect. Immun.* **61**(12), 5090–6.

54. Czuprynski, C. J. *et al.* (1994). Administration of anti-granulocyte mAb RB6-8C5 impairs the resistance of mice to *Listeria monocytogenes* infection. *J. Immunol.* **152**(4), 1836–46.

55. Rosen, H., S. Gordon, and R. J. North (1989). Exacerbation of murine listeriosis by a monoclonal antibody specific for the type 3 complement receptor of myelomonocytic cells. Absence of monocytes at infective foci allows *Listeria* to multiply in nonphagocytic cells. *J. Exp. Med.* **170**(1), 27–37.

56. Fang, F. C. (2004). Antimicrobial reactive oxygen and nitrogen species: concepts and controversies. *Nat. Rev. Microbiol.* **2**(10), 820–32.

57. Amer, A. O. and M. S. Swanson (2002). A phagosome of one's own: a microbial guide to life in the macrophage. *Curr. Opin. Microbiol.* **5**(1), 56–61.

58. Shiloh, M. U. *et al.* (1999). Phenotype of mice and macrophages deficient in both phagocyte oxidase and inducible nitric oxide synthase. *Immunity* **10**(1), 29–38.

59. Serbina, N. V. *et al.* (2003). TNF/iNOS-producing dendritic cells mediate innate immune defense against bacterial infection. *Immunity* **19**(1), 59–70.

60. Endres, R. *et al.* (1997). Listeriosis in p47(phox$^{-/-}$) and TRp55$^{-/-}$ mice: protection despite absence of ROI and susceptibility despite presence of RNI. *Immunity* **7**(3), 419–32.

61. Dinauer, M. C., M. B. Deck, and E. R. Unanue (1997). Mice lacking reduced nicotinamide adenine dinucleotide phosphate oxidase activity show increased susceptibility to early infection with *Listeria monocytogenes. J. Immunol.* **158**(12), 5581–3.

62. Kuziel, W. A. *et al.* (1997). Severe reduction in leukocyte adhesion and monocyte extravasation in mice deficient in CC chemokine receptor 2. *Proc. Natl Acad. Sci. U S A* **94**(22), 12053–8.

63. Kurihara, T. *et al.* (1997). Defects in macrophage recruitment and host defense in mice lacking the CCR2 chemokine receptor. *J. Exp. Med.* **186**(10), 1757–62.

64. Sato, N. *et al.* (2000). CC chemokine receptor (CCR)2 is required for Langerhans cell migration and localization of T helper cell type 1 (Th1)-inducing dendritic cells. Absence of CCR2 shifts the *Leishmania* major-resistant phenotype to a susceptible state dominated by Th2 cytokines, b cell outgrowth, and sustained neutrophilic inflammation. *J. Exp. Med.* **192**(2), 205–18.

65. Izikson, L. *et al.* (2000). Resistance to experimental autoimmune encephalomyelitis in mice lacking the CC chemokine receptor (CCR)2. *J. Exp. Med.* **192**(7), 1075–80.

66. Vecchi, A. *et al.* (1999). Differential responsiveness to constitutive vs. inducible chemokines of immature and mature mouse dendritic cells. *J. Leukoc. Biol.* **66**(3), 489–94.

67. Mack, M. *et al.* (2001). Expression and characterization of the chemokine receptors CCR2 and CCR5 in mice. *J. Immunol.* **166**(7), 4697–704.

68. Merad, M. *et al.* (2002). Langerhans cells renew in the skin throughout life under steady-state conditions. *Nat. Immunol.* **3**(12), 1135–41.

69. Geissmann, F., S. Jung, and D. R. Littman (2003). Blood monocytes consist of two principal subsets with distinct migratory properties. *Immunity* **19**(1), 71–82.

70. Peters, W. *et al.* (2004). CCR2-dependent trafficking of F4/80dim macrophages and CD11cdim/intermediate dendritic cells is crucial for T cell recruitment to lungs infected with *Mycobacterium tuberculosis*. *J. Immunol.* **172**(12), 7647–53.

71. Kurihara, T. and R. Bravo (1996). Cloning and functional expression of mCCR2, a murine receptor for the C–C chemokines JE and FIC. *J. Biol. Chem.* **271**(20), 11603–7.

72. Sozzani, S. *et al.* (1997). Receptor expression and responsiveness of human dendritic cells to a defined set of CC and CXC chemokines. *J. Immunol.* **159**(4), 1993–2000.

73. Sheffler, L. A. *et al.* (1995). Exogenous nitric oxide regulates IFN-gamma plus lipopolysaccharide-induced nitric oxide synthase expression in mouse macrophages. *J. Immunol.* **155**(2), 886–94.

74. Lu, L. *et al.* (1996). Induction of nitric oxide synthase in mouse dendritic cells by IFN-gamma, endotoxin, and interaction with allogeneic

T cells: nitric oxide production is associated with dendritic cell apoptosis. *J. Immunol.* **157**(8), 3577−86.

75. Albina, J. E. and W. L. Henry, Jr. (1991). Suppression of lymphocyte proliferation through the nitric oxide synthesizing pathway. *J. Surg. Res.* **50**(4), 403−9.

76. Eriksson, S., B. J. Chambers, and M. Rhen (2003). Nitric oxide produced by murine dendritic cells is cytotoxic for intracellular *Salmonella enterica sv. Typhimurium*. *Scand. J. Immunol.* **58**(5), 493−502.

77. Cheminay, C., A. Mohlenbrink, and M. Hensel (2005). Intracellular *Salmonella* inhibit antigen presentation by dendritic cells. *J. Immunol.* **174**(5), 2892−9.

78. Fremond, C. M. *et al.* (2004). Fatal *Mycobacterium tuberculosis* infection despite adaptive immune response in the absence of MyD88. *J. Clin. Invest.* **114**(12), 1790−9.

79. Bodnar, K. A., N. V. Serbina, and J. L. Flynn (2001). Fate of *Mycobacterium tuberculosis* within murine dendritic cells. *Infect. Immun.* **69**(2), 800−9.

80. Buettner, M. *et al.* (2005). Inverse correlation of maturity and antibacterial activity in human dendritic cells. *J. Immunol.* **174**(7), 4203−9.

81. Kolb-Maurer, A. *et al.* (2000). *Listeria monocytogenes*-infected human dendritic cells: uptake and host cell response. *Infect. Immun.* **68**(6), 3680−8.

82. Paschen, A. *et al.* (2000). Human dendritic cells infected by *Listeria monocytogenes*: induction of maturation, requirements for phagolysosomal escape and antigen presentation capacity. *Eur. J. Immunol.* **30**(12), 3447−56.

83. Guzman, C. A. *et al.* (1995). Interaction of *Listeria monocytogenes* with mouse dendritic cells. *Infect. Immun.* **63**(9), 3665−73.

84. Neild, A. L. and C. R. Roy (2003). *Legionella* reveal dendritic cell functions that facilitate selection of antigens for MHC class II presentation. *Immunity* **18**(6), 813−23.

85. Macpherson, A. J. and T. Uhr (2004). Induction of protective IgA by intestinal dendritic cells carrying commensal bacteria. *Science* **303**(5664), 1662−5.

86. Hopkins, S. A. *et al.* (2000). A recombinant *Salmonella typhimurium* vaccine strain is taken up and survives within murine Peyer's patch dendritic cells. *Cell Microbiol.* **2**(1), 59−68.

87. Brzoza, K. L., A. B. Rockel, and E. M. Hiltbold (2004). Cytoplasmic entry of *Listeria monocytogenes* enhances dendritic cell maturation and T cell differentiation and function. *J. Immunol.* **173**(4), 2641−51.

88. Feng, H. *et al.* (2005). *Listeria*-infected myeloid dendritic cells produce IFN-beta, priming T cell activation. *J. Immunol.* **175**(1), 421—32.

89. Kolb-Maurer, A. *et al.* (2003). Production of IL-12 and IL-18 in human dendritic cells upon infection by *Listeria monocytogenes. FEMS Immunol. Med. Microbiol.* **35**(3), 255—62.

90. Takeda, K. and S. Akira (2004). TLR signaling pathways. *Semin. Immunol.* **16**(1), 3—9.

91. Seki, E. *et al.* (2002). Critical roles of myeloid differentiation factor 88-dependent proinflammatory cytokine release in early phase clearance of *Listeria monocytogenes* in mice. *J. Immunol.* **169**(7), 3863—8.

92. Edelson, B. T. and E. R. Unanue (2002). MyD88-dependent but Toll-like receptor 2-independent innate immunity to *Listeria*: no role for either in macrophage listericidal activity. *J. Immunol.* **169**(7), 3869—75.

93. Adachi, O. *et al.* (1998). Targeted disruption of the MyD88 gene results in loss of IL-1- and IL-18-mediated function. *Immunity* **9**(1), 143—50.

94. Tsuji, N. M. *et al.* (2004). Roles of caspase-1 in *Listeria* infection in mice. *Int. Immunol.* **16**(2), 335—43.

95. Hayashi, F. *et al.* (2001). The innate immune response to bacterial flagellin is mediated by Toll-like receptor 5. *Nature* **410**(6832), 1099—103.

96. Ito, S. *et al.* (2005). CpG oligodeoxynucleotides enhance neonatal resistance to *Listeria* infection. *J. Immunol.* **174**(2), 777—82.

97. Inohara, N. *et al.* (2003). Host recognition of bacterial muramyl dipeptide mediated through NOD2. Implications for Crohn's disease. *J. Biol. Chem.* **278**(8), 5509—12.

98. Girardin, S. E. *et al.* (2003). Nod2 is a general sensor of peptidoglycan through muramyl dipeptide (MDP) detection. *J. Biol. Chem.* **278**(11), 8869—72.

99. Li, J. *et al.* (2004). Regulation of IL-8 and IL-1beta expression in Crohn's disease associated NOD2/CARD15 mutations. *Hum. Mol. Genet.* **13**(16), 1715—25.

100. Girardin, S. E. *et al.* (2003). Nod1 detects a unique muropeptide from Gram-negative bacterial peptidoglycan. *Science* **300**(5625), 1584—7.

101. Kobayashi, K. S. *et al.* (2005). Nod2-dependent regulation of innate and adaptive immunity in the intestinal tract. *Science* **307**(5710), 731 4.

102. Chin, A. I. *et al.* (2002). Involvement of receptor-interacting protein 2 in innate and adaptive immune responses. *Nature* **416**(6877), 190—4.

103. Kobayashi, K. *et al.* (2002). RICK/Rip2/CARDIAK mediates signalling for receptors of the innate and adaptive immune systems. *Nature* **416**(6877), 194—9.

104. Hauf, N. *et al.* (1997). *Listeria monocytogenes* infection of P388D1 macro-phages results in a biphasic NF-kappaB (RelA/p50) activation induced by lipoteichoic acid and bacterial phospholipases and mediated by IkappaBalpha and IkappaBbeta degradation. *Proc. Natl Acad. Sci. U S A* **94**(17), 9394−9.

105. O'Connell, R. M. *et al.* (2005). Immune activation of type I IFNs by *Listeria monocytogenes* occurs independently of TLR4, TLR2, and receptor interacting protein 2 but involves TNFR-associated NF kappa B kinase-binding kinase 1. *J. Immunol.* **174**(3), 1602−7.

106. Pron, B. *et al.* (2001). Dendritic cells are early cellular targets of *Listeria monocytogenes* after intestinal delivery and are involved in bacterial spread in the host. *Cell Microbiol.* **3**(5), 331−40.

107. Vollstedt, S. *et al.* (2003). Flt3 ligand-treated neonatal mice have increased innate immunity against intracellular pathogens and efficiently control virus infections. *J. Exp. Med.* **197**(5), 575−84.

108. Jung, S. *et al.* (2002). In vivo depletion of CD11c[(+)] dendritic cells abrogates priming of CD8[(+)] T cells by exogenous cell-associated anti-gens. *Immunity* **17**(2), 211−20.

109. Kolb-Maurer, A. *et al.* (2003). Induction of IL-12 and IL-18 in human dendritic cells upon infection by *Listeria monocytogenes*. *FEMS Immunol. Med. Microbiol.* **35**(3), 255−62.

110. Alanir, R. C. *et al.* (2004). Increased dendritic cell numbers impair protective immunity to intracellular bacteria despite augmenting antigen-specific CD8[+] T lymphocyte responses. *J. Immunol.* **172**, 3725−35.

CHAPTER 6

Interactions between natural killer and dendritic cells during bacterial infections

Guido Ferlazzo

Istituto Nazionale Ricerca sul Cancro and University of Messina

Natural killer (NK) cells represent a distinct lymphoid population characterized by unique phenotypic and functional features. NK cells were originally identified on a functional basis as this denomination was assigned to lymphoid cells capable of lysing tumor cell lines in the absence of prior stimulation in vivo or in vitro[1]. Both their origin and the mechanism(s) mediating their function remained mysterious until recently. Regarding their origin, it has been shown that NK cells derive from a precursor common to T cells and expressing the $CD34^+CD7^+$ phenotype. In addition, functional NK cells can be obtained in vitro and in vivo from ($CD34^+$) haematopoietic precursors isolated from several different sources[2–6]. The cell maturation in vitro has been shown to require appropriate feeder cells and/or IL-15. The molecular mechanisms underlying the ability of NK cells to discriminate between normal and tumor cells, predicted by the "missing self hypothesis"[7]. have been clarified only during the past decade. It has been shown that NK cells recognize MHC-class I molecules through surface receptors delivering inhibitory, rather than activating, signals. Accordingly, NK cells lyse target cells that have lost (or express low amounts of) MHC class I molecules. This event occurs frequently in tumors or in cells infected by some viruses such as certain herpesviruses or adenoviruses. In addition to providing a first line of defence against viruses, NK cells release various cytokines and chemokines. These released cytokines can control bacterial spreading but also induce or modulate inflammatory responses, hematopoiesis, and control the growth and function of monocytes and granulocytes. Finally, the functional links between NK and dendritic cells (DCs) have been widely investigated in recent years and different studies have demonstrated that reciprocal activations ensue upon NK/DC interactions.

More recently, the anatomical sites where these interactions take place have been identified together with the related cell subsets involved[8]. Remarkably, there is now "in vivo" evidence that this cellular cross-talk occurring during the innate phase of the immune response against bacteria or bacterial products can deeply affect the magnitude and the quality of the subsequent adaptive response[9].

These new experimental evidences emphasize the relevance of the interplay between DCs and NK cells during bacterial infections.

6.1 NK CELLS EVOLVED TO COOPERATE WITH THE ADAPTIVE IMMUNITY

Thus, NK cells are not merely cytotoxic lymphocytes competent in containing viral and tumor spreading but can now rather be considered as crucial fine-tuning effector cells widely involved in different phases of the immune response.

Indeed, while NK cells have long been defined as "primitive" and "non-specific" effector cells, we have now a different perception of these cells. First, it is evident that NK cells evolved to adapt to and cooperate with mechanisms of the specific immunity: they have evolved receptors for the Fc portion of IgG that allow killing of antibody-coated target cells or certain pathogens; they also release a number of cytokines that regulate T cell activation and function. Importantly, an early activation of NK cells during immune responses may influence the quality of the subsequent T cell response by inducing a Th1 polarization. Second, NK cells have evolved a mechanism allowing the rapid detection and killing of potentially dangerous cells characterized by an altered expression of MHC class I antigens due to infections. Human NK cells have been shown to express different human leukocyte antigen (HLA)-class I-specific inhibitory receptors. A family of these receptors (termed killer Ig-like receptors (KIR)) detect shared allelic determinants of HLA class I molecules while others display a more "promiscuous" pattern of recognition and are characterized by a broad specificity for different HLA class I molecules (LIR1/ILT2) or recognize the HLA-class Ib HLA-E molecules (CD94/NKG2A)[10–14].

This mechanism is sophisticated (KIRs detect allelic determinants of HLA class I molecules) and of recent evolution since murine NK cells lack KIRs (a similar function is mediated by structurally different receptors[10] and major differences in the type and specificity of expressed KIRs exist in chimpanzees, a species that diverged from humans only

5 million years ago[15]. This clearly means that KIRs have evolved recently, paralleling the rapid evolution of HLA-class I molecules.

6.2 CYTOTOXIC NK CELLS AND CYTOKINE SECRETING NK CELLS

Although NK cells were initially described as cytotoxic effectors in peripheral blood and were found to lyse tumor cells without prior activation, recent studies suggest that a possibly immunoregulatory subset of NK cells responds to activation mainly with cytokine secretion. This latter subset should play the prominent protective role during bacterial infections. For instance, IFNγ, a major cytokine released by activated NK cells, represents the principal phagocyte-activating factor[5], indicating the crucial function of NK cell activation during bacterial infections.

Human peripheral blood mononuclear cells contain around 10% of NK cells[16]. The majority ($\geq$95%) belongs to the CD56^{dim}CD16$^+$ cytolytic NK subset[17–19]. These cells carry homing markers for inflamed peripheral sites and carry perforin to rapidly mediate cytotoxicity[17,18]. The minor NK subset in blood ($\leq$5%) is CD56^{bright}CD16$^-$ cells[17–19]. These NK cells lack perforin (or have low levels of it), but secrete large amounts of IFNγ and TNFβ upon activation and are superior to CD56^{dim} NK cells in these functions[18,19]. In addition, they display homing markers for secondary lymphoid organs, namely CCR7 and CD62L[7].

Another difference between these subsets can be found in their receptors mediating target recognition. Human NK cell recognition of target cells is guided by the balance of activating and inhibitory signals given by different groups of surface receptors. The main activating receptors constitutively found on all NK cells in peripheral blood are NKG2D and the nitrogen catabolite repressors (NCRs) NKp30 and NKp46[20]. All of them probably recognize molecules that are upregulated upon cellular stress[21,22]. However, only the stress-induced NKG2D ligands MICA/B and ULBPs have so far been identified[23,24]. While cytotoxic peripheral blood CD56^{dim} NK cells are able to target antibody-opsonized cells via their low affinity FcγRIII/CD16 molecule, immunoregulatory CD56^{bright} NK cells lack this receptor nearly entirely[18,19].

Most inhibitory NK cell receptors engage MHC class I molecules on target cells[12]. They can be distinguished into two groups, detecting either common allelic determinants of MHC class I, or MHC class I expression in general. The KIR receptors constituting the first group distinguish polymorphic HLA-A, -B and -C molecules. The inhibitory receptor surveying MHC class I expression in general are more heterogeneous. They include the

LIR1/ILT2 molecule with a broad specificity for different HLA class I molecules and the CD94/NKG2A heterodimer, specific for HLA-E whose surface expression is dictated by the availability of HLA class I heavy chain signal peptides. The MHC class I allele specific KIR receptors are expressed on subsets of $CD56^{dim}CD16^+$ cytolytic NK cells, while the immunoregulatory $CD56^{bright}CD16^-$ NK subset expresses uniformly CD94/NKG2A and lacks KIRs[18].

These phenotypes are consistent with the hypothesis that $CD56^{dim}CD16^+$ NK cells are terminally differentiated effectors that carry the whole panel of sophisticated activating and inhibitory receptors to even detect allelic HLA loss and can readily lyse aberrant cells at peripheral inflammation sites. On the contrary, $CD56^{bright}CD16^-$ NK cells might perform an immunoregulatory function in secondary lymphoid tissues and release large amount of effector cytokines able to control pathogen spreading.

6.3 NK CELLS IN BLOOD AND IN SECONDARY LYMPHOID ORGANS

Peripheral blood is the most accessible source of human NK cells. Therefore, most studies have been performed with NK cell populations from this organ and it was assumed that most NK cells circulate in human blood after their emigration from the bone marrow. Recently, however, it has been shown that a substantial amount of human NK cells homes to secondary lymphoid organs. These amount to around 5% of mononuclear cells in uninflamed lymph nodes and 0.4–1% in inflamed tonsils and lymph nodes[25,26]. These NK cells constitute a remarkable pool of innate effector cells, since lymph nodes harbor 40% of all lymphocytes, while peripheral blood contains only 2% of all lymphocytes[27,28]. Therefore lymph node NK cells are in the absence of infection and inflammation 10 times more abundant than blood NK cells.

As expected from CCR7 and CD62L expression on $CD56^{bright}$ NK cells in blood, the $CD56^{bright}$ NK subset is enriched in all secondary lymphoid organs analysed so far (lymph nodes, tonsils and spleen)[26]. In spleen, around 15% and in tonsils and lymph nodes around 75% of NK cells belong to the $CD56^{bright}$ subset. Lymph node and tonsil NK cells, however, uniformly lack FcγRIII/CD16 and KIRs, while the $CD56^{dim}$ NK cells in spleen express these molecules[26]. Surprisingly and in contrast to $CD56^{bright}$ blood NK cells, lymph node and tonsil NK cells show no or very low expression of the constitutive NCRs NKp30 and NKp46[25,26]. On the other hand, the inducible NCR NKp44, undetectable on NK cells directly isolated from peripheral blood, is upregulated on NK cells harbored in inflamed tonsils[26]. Therefore, target cell recognition by lymph node and tonsil NK cells seems to be mainly influenced by NKG2D as activating and CD94/NKG2A as inhibitory receptor[25,26]. This NK receptor repertoire of lymph node and tonsil NK cells is similar to CD56^{bright} blood NK cells, but even more restricted by the absence of constitutive NCRs.

With respect to NK function, secondary lymphoid tissue NK cells show an impressive plasticity. Although lymph node and tonsil NK cells are initially perforin negative and show no cytolytic activity against MHC class I low and ULBP^{high} targets, perforin and cytotoxicity can be upregulated by IL-2 within 3–7 days[26]. Interestingly, NK cells from secondary lymphoid organs gain during the same time-period expression of CD16, NCRs NKp30, NKp46 and NKp44 as well as KIRs[26]. Therefore, activation converts lymph node and tonsil NK cells into effectors, similar to the terminally differentiated CD56^{dim}CD16$^+$ blood NK cells with cytolytic function and the sophisticated set of inhibitory and activating receptors.

6.4 DENDRITIC CELLS AS EARLY ACTIVATORS OF NK CELL FUNCTIONS

The mentioned human NK cell compartments are probably all involved in early innate immune responses. Recent studies have demonstrated that during the innate phase of the immune response NK cells can also mediate DC maturation[29–31]. This activation is not unidirectional, because the interaction between mature, but not immature, DCs and NK cells results in NK cell proliferation, IFNγ production and induction of cytolytic activity[8,32,33]. Thus, DCs have now emerged as the activators of NK response in the early phases of the immune response, i.e. before an adaptive immune response had been evoked and T cell derived cytokine, such as IL-2, could be produced.

Remarkably, DC-induced NK cell cytolytic activity was directed not only toward tumor cells, but also against immature DCs (iDCs). The NK-mediated killing of DCs was mostly dependent on the NKp30 NCR. Other activating receptors or co-receptors played virtually no role[33]. This would imply that DCs express the ligand for NKp30, but not for other major triggering NK receptors.

During NK activation by myeloid DCs both soluble factors as well as cell-to-cell contact seem to be important.

In mice, induction of NK cell cytotoxicity was entirely blocked by transwell separation of DCs and NK cells, indicative for a major contribution of

DC surface receptors in NK activation[32]. In addition, DC derived IFNα/β, IL-12 and IL-18 have been reported to be crucial in murine NK activation[34–36]. While IL-12 was mainly implicated in NK mediated INFγ secretion, IFNα/β seems required for cytotoxicity of NK cells[35,37]. In addition, secretion of IL-2 by DCs stimulated with microbial stimuli might also contribute to NK activation[38,39].

In humans, NK activation by DCs was not significantly disrupted by trans-well separation of the two cell types, indicating a major contribution of soluble factors in NK activation[40,41]. In one study, DCs were unable to activate NK cells in the presence of neutralizing antibodies for IL-12 and IL-18[41]. Other studies demonstrated that NK activation by DC subsets correlates with IL-12 secretion by these DCs, while IL-15 and IL-18 secretion were not indicative for NK activation[42,43]. Therefore, IL-12 might play an important role in human NK activation by DCs.

Apart from myeloid DCs that efficiently activate human and mouse NK cells, plasmacytoid DCs (pDCs) might also contribute to NK activation. This DC subset has been found to produce 200 to 1000 times more type I interferons than other blood cells after viral challenge[44] and IFN$\alpha\beta$ are critical cytokines for inducing NK cell-mediated lysis of virus-infected targets[45–47]. IFN$\alpha\beta$ secretion seems, however, not only the signature of pDCs, but also myeloid DCs can secrete substantial amounts of these cytokines upon direct viral infection[48], as well as upon bacterial infection[49]. Therefore, the contributions of myeloid and plasmacytoid DCs in type I interferon mediated NK activation remains to be established.

While cytokines play a dominant role in human NK cell activation by DCs, IFNα treatment of DCs in addition upregulates the NKG2D ligands MICA/B on monocyte-derived DCs and these molecules seem to activate resting NK cells in a cell contact dependent manner[50]. MICA/B upregulation on DCs upon IFNα exposure seems to be mediated by DC derived autocrine/paracrine IL-15[51]. Therefore, cell-contact might contribute under certain inflammatory conditions to NK activation by human DCs.

6.5 EFFECT OF BACTERIAL INFECTIONS ON DC/NK CELL INTERACTIONS

It is conceivable that DC/NK cell interactions may occur primarily during infections. Therefore it was important to analyze the effect of live bacteria on the cross-talk between DCs and NK cells. Two different models of bacterial infection have been analyzed[42]. One represented by the extracellular bacteria *Escherichia coli* (*E. coli*), the other by the intracellular mycobacterium BCG, capable of efficiently infecting DCs[52,53]. In both systems, bacterial infection of DCs led to a particularly rapid NK cell activation.

6.5.1 Autologous DC-induced proliferation of NK cells is enhanced by BCG

Since mature DCs are capable of inducing proliferation of autologous NK cells, [33] it was then further investigated whether the infection of DCs with BCG had any effect on this capability. BCG was employed as infective agent because of its ability to efficiently infect DCs without undergoing substantial proliferation (thus not interfering with 3H-thymidine incorporation assays). In this study, it was confirmed that DCs, derived from monocytes in the presence of GM-CSF and IL-4, were able to induce NK cells proliferation. It is of note that both the NK cell proliferation and the number of viable NK cells recovered after 5 days of culture were significantly increased in the presence of BCG. Control experiments with culture containing NK cells and BCG alone did not lead to NK cell proliferation.

The observed NK cell proliferation is likely to be sustained by lymphokines such as IL-2 and IL-15[38,54,55]. Although these cytokines were detected at extremely low levels in the supernatants derived from DCs cultured in the presence of bacteria, it is not possible to exclude their role in DC-mediated NK cell expansion. Indeed, a more recent study indicate a relevant role of the membrane-bound form of IL-15 on human DCs stimulated by different inflammatory stimuli, including LPS, in DC-dependent NK cell proliferation. Notably, in this experimental model, CD56[bright] lymph node NK cells were preferentially expanded during DC/NK coculture[8].

6.5.2 Effectiveness of BCG and *E. coli* in inducing an activated NK cell phenotype

Previous studies have shown that, following stimulation with LPS, DCs induce the expression, in NK cells, of the early activation marker CD69[29,31]. Accordingly, DCs that had been exposed to living bacteria could also induce activation markers on NK cells. Both CD69 and HLA-DR were also upregulated in the presence of bacteria. It is of note that whereas CD69 and HLA-DR can be detected after interaction of NK cells with non-infected immature DCs, DC infection resulted in a greater increase of the expression of both molecules on NK cell surface. In addition, a de novo expression of CD25 could be detected. This de novo expression may be functionally relevant since the expression of CD25, a component of the IL-2 receptor complex, renders NK cells highly responsive to IL-2. Neither BCG nor *E. coli* could induce the expression of these activation markers in the absence of DCs.

Since the NKp30 receptor is primarily involved in DCs recognition and lysis by NK cells[33], it was also analyzed whether mAb-mediated masking of NKP30 could interfere with the NK cell activation induced by DCs and bacteria. The addition of anti-NKp30 mAb did not modify the expression of CD69, HLA-DR and CD25. These data clearly indicate that the induction of an activated phenotype in NK cells is not mediated via NKp30. Therefore, this triggering receptor, which plays a major role in the recognition and lysis of DCs, does not appear to play any substantial role in the activating signal delivered by DCs to NK cells.

6.5.3 DC editing by NK cells activated by DCs and bacteria

Activated NK cells can lyse autologous iDCs while they are less effective against mature DCs[33,56]. In addition, a short term coculture of resting NK cells with DCs that had been pulsed with LPS or heat-killed *Mycobacterium tuberculosis*, has been reported to result in increase of NK-mediated cytotoxicity against the Daudi target cell line[29]. It was further investigated whether NK cells cocultured with DCs that had been infected with live bacteria, could lyse autologous iDCs. After 24 h of coculture, only NK cells cultured in the presence of infected DCs could lyse autologous iDCs. After this time interval, uninfected iDCs failed to induce NK cell cytotoxicity. Remarkably, infected DCs were less susceptible to the lysis, as compared to iDCs cultured in the absence of bacteria. These results were obtained after as few as 24 h of NK/DC coculture.

In the same set of experiments it was also analyzed whether polyclonal NK cell lines cultured for over 1 week in IL-2 could discriminate between infected and non-infected autologous DCs. IL-2-cultured NK cells lysed uninfected iDCs very efficiently, but were less effective against infected DCs. Therefore, the resistance of infected DCs to NK-mediated lysis does not depend upon the degree of NK cell activation, but rather reflects an intrinsic property of infected DCs themselves.

In this context, the arrival of NK cells to inflamed tissues and their encounter with iDCs may appear paradoxical, as it would lead to depletion of antigen-presenting cells (APCs). Nevertheless, DCs exposed to bacteria become highly resistant to NK cell-mediated lysis. Therefore, while they can positively influence NK cells, they are not damaged by NK cells themselves and are allowed to migrate to secondary lymphoid organs.

In turn, NK cells undergo both activation and proliferation, display a rapid increase in their cytolytic activity and may release large amounts of cytokines, including TNF-α, GM-CSF and IFNγ[54], which may further amplify the inflammatory response.

6.5.4 The resistance of infected DCs to NK-mediated cytotoxicity is due to the upregulation of HLA class I molecules

The different susceptibility of iDCs versus mature DCs is primarily related to differences in surface expression of HLA class I molecules[57]. In order to determine whether the early resistance of infected DCs to NK-mediated lysis reflected a rapid upregulation of HLA class I molecules, the surface density of HLA class I on DCs, cultured alone or in the presence of either BCG or *E. coli*, was comparatively analyzed on a quantitative basis. Cell size and the expression of HLA class I molecules were evaluated by flow cytometry. By the simultaneous evaluation of cell size and fluorescence intensity the number of HLA class I molecules/μ^2 of DCs cell surface could be calculated. Two days (48 h) after infection, DCs increased the number of HLA class I molecules/μ^2 approximately ten-fold with both BCG and *E. coli*. In view of the high expression of HLA class I in infected DCs, it appears conceivable that resistance to NK-mediated lysis could be a distinct consequence of this phenomenon. On the other hand, another possible explanation could be that BCG and *E. coli* could down-regulate the expression of the ligands for triggering receptors of NK cells. In order to discriminate between these two possibilities, cytolytic tests in the presence of mAbs (IgM isotype) specific for HLA class I molecules were performed. Upon mAb-mediated masking of HLA class I molecules, a sharp increase of cytolytic activity against infected DCs could be detected. This clearly indicates that the resistance of infected DCs to NK-mediated lysis is due to increased inhibitory interactions occurring between HLA class I and inhibitory receptors expressed on NK cells. The anti-HLA class I mAb-induced restoration of the cytolytic activity and the inhibition of this activity by anti-NKp30 mAb, manifestly pointed out that infected DCs also express levels of ligands for this triggering NK receptor sufficient to induce NK cell activation.

A relevant question related to the above-presented data is why bacterial infection should lead to a rapid induction of NK cell activation and cytotoxicity. Thus, while NK cell recruitment and activation result in production of cytokines and chemokines which may contribute to the defense against bacterial spreading, the cytolytic activity of NK cells does not exert any direct effect against bacteria. In this regard, a physiopathologic mechanism, in which activated NK cells could play a role in the homeostasis of the immune response during bacterial infections, has been proposed[33,58]. This model is based on the evidence that activated NK cells are inefficient in killing infected DCs (as discussed above) but they can efficiently lyse uninfected DCs. Accordingly, the presence of activated NK cells in tissues and lymph nodes, inflamed because of bacterial infection, may limit an overwhelming recruitment of iDCs at a stage in which bacteria have already been eliminated and infection has been controlled. Therefore, the ability of NK cells to discriminate between infected and uninfected DCs may suggests that NK cells play an important regulatory role by selectively editing APC during bacterial infection and thus switching off an excessive immune response. This mechanism may be particularly useful in preventing tissue damage.

Exposure of DCs to bacteria rapidly induce DC maturation and expression of functionally important surface molecules, including CD80 and CD86 coreceptors, HLA molecules and CCR7[42]. Thus, DCs acquire rapidly the ability to efficiently function as professional APC and to migrate to secondary lymphoid organs, where they can interact with T cells and evoke a prompt adaptive immune response against infecting bacteria. In addition, DCs that had encountered bacteria can also rapidly and markedly potentiate an important effector arm of the innate immunity by inducing a rapid activation of NK cells.

A conceivable question may now be how and where DCs and NK cells can meet each other.

6.6 SITES OF NK/DC INTERACTION DURING BACTERIAL INFECTION

6.6.1 DC/NK cell cross-talk at sites of bacterial invasion

The recent data on NK cell activation by DCs suggest that these APCs initiate the early and innate NK activation during the immune response. The question remains as to where this interaction takes place. No direct evidence exists that DCs and NK cells encounter at sites of bacterial infection. Nevertheless, microbial invasion causes tissue inflammation and one site of DC/NK interaction is inflamed tissue. NK cells have been found in close contact to DCs in lesions of allergen induced atopic eczema/dermatitis syndrome[59]. Moreover, the chemokine receptor repertoire and the chemokine responsiveness of CD56^{dim}CD16$^+$ blood NK cells suggest that they can home to sites of inflammation efficiently[17]. The cytotoxic blood NK subset migrates efficiently in response to IL-8 and soluble fractalkine and expresses the respective receptors for these chemokines, CXCR1 and CX3CR1. Both chemokines are induced by proinflammatory cytokines like IL-1 and TNFα[60,61]. Fractalkine mediates adhesion to endothelia and emigration of NK cells from the blood stream, while IL-8 mediates further migration to the site of inflammation[62]. Therefore, cytotoxic CD56^{dim}CD16$^+$ blood NK cells are able to home to inflamed tissues where they can encounter dendritic cells, which are resident in peripheral tissue sites.

The DC/NK encounter at sites of inflammation can either result in DC maturation by modest NK infiltration or in immature DC lysis due to large NK cell infiltrates[31]. The maturation of DCs upon NK encounter has been largely attributed to TNFα secretion by NK cells[29,31]. Both NK effector mechanisms will deplete the inflamed tissue of DCs either by maturation-induced migration or by killing. This will deprive DC trophic bacteria of their host cells at the site of infection.

6.6.2 DCs infected by bacteria activate and induce maturation of NK cells in secondary lymphoid organs

Until recently only limited information had been available on NK cells located in lymphoid tissues, and therefore NK cells have mainly been considered as effector cells harbored in the blood stream and able to promptly extravasate to inflamed tissues. The evidence that a large amount of NK cells is located in uninflamed lymph nodes[26] suggests secondary lymphoid organs as important sites of NK cell activation. Indeed, CD56^{bright}CD16$^-$ NK cells isolated from uninflamed human lymph nodes become strongly cytolytic upon stimulation with IL-2. The de novo acquired cytotoxic properties were accompanied by the expression of both activating and inhibitory receptors.

In addition, perforin negative NK cells located in secondary lymphoid organs might play different roles prior to maturation into cytolytic effectors, such as secretion of critical immunoregulatory cytokines upon activation. In this regard, it has been demonstrated that peripheral blood CD56^{bright}CD16$^-$ NK cells produce significantly higher levels of cytokines than their CD56^{dim}CD16$^+$ counterpart[19,54]. Similarly, the CD56^{bright}CD16$^-$ NK subset located in secondary lymphoid organs produces relevant cytokines prior to maturation into cytolytic effector cells[26].

Interestingly, several reports have recently shown that DCs elicit IFNγ secretion by autologous NK cells[29–33]. Specifically, NK cells from secondary lymphoid organs are particularly effective in this function and when they are cocultured with autologous DCs stimulated by bacterial products can produce IFNγ within 6 h[8]. This is of interest since IFNγ represents a main cytokine for phagocyte activation. Indeed, CD56[bright] CD16[−] NK cells produce significantly higher levels of IFNγ, TNFβ, GM-CSF, IL-10 and IL-13 protein in response to monokines produced by DCs, such as IL-12, IL-15, IL-18 and IL-1β[19]. Based on these findings, NK cells in secondary lymphoid organs should not be referred to as merely "immature" NK cells but rather as effector cells whose functional plasticity enables them to accomplish different sequential tasks during immune responses.

Since DCs mature and migrate to secondary lymphoid tissues following an encounter with bacteria, they might encounter CD56[bright] CD16[−] NK cells there in the very early phase of an immune response prior to T cell activation. As discussed in detail below, this could result in local cytokine release by NK cells, which might be able to shape the following adaptive immune response and probably also APC functions[63].

6.7 ROLES FOR NK CELLS IN DC-MEDIATED T CELL POLARIZATION

We discussed above that a recent "in vitro" model proposed that NK cells might play an important regulatory role by selectively editing APC during the course of immune responses. NK-mediated lysis of immature, but not mature DCs might select an immunogenic DC population during the initiation of immune responses.

In addition to removal of non-immunogenic DCs, NK cells also secrete IFNγ upon encounter with DCs[33]. As a consequence, subsequent T cell polarization may be influenced. Indeed, in vitro studies in both mouse and human systems have demonstrated the importance of IFNγ in the polarization of type 1 immune response. Interestingly, in a murine model of skin graft rejection, the recognition of donor DCs by host NK cells led to modulation of Th1/Th2 cell development. Namely, turning host NK cells off was sufficient to skew the alloresponse to Th2[64].

Thus, as already mentioned, the encounter of mature DCs with perforin[neg] NK cells located in secondary lymphoid organs should lead to a critical immunoregulatory role[65], as DCs can induce NK cells of human secondary lymphoid organs to secrete IFNγ[8]. Moreover, DCs selectively

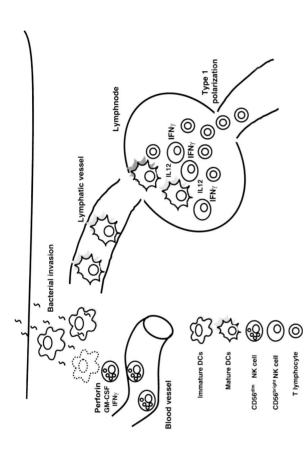

Figure 6.1. During bacterial infection DCs and NK cells can interact in tissues inflamed by bacterial invasion and in secondary lymphoid organs. Upon bacterial infection, NK cells get recruited from peripheral blood to inflamed tissues (upper left). Cytolytic, perforin$^+$CD56^{dim} NK cells, which constitute the majority of peripheral blood NK cells, carry the necessary chemokine receptors to home to sites of inflammation. They can edit DCs here before these cells mature and migrate to secondary lymphoid organs. In addition, upon encounter with DCs, NK cells can release cytokines that can control bacterial spreading but also induce or modulate inflammatory responses, as well as control the growth and function of monocytes/macrophages, granulocytes and DCs themselves. In secondary lymphoid organs like lymph nodes, DCs can meet with regulatory CD56^{bright} NK cells (lower right), which constitute the majority of NK cells in secondary lymphoid tissues. The interaction between DCs matured by bacteria and CD56^{bright} NK cells ensure a large release of IFNγ by NK cells via DC-derived IL-12. The realease of IFNγ and/or other immunomodulatory cytokines influence the magnitude and the polarization of the DC-initiated adaptive immune response.

stimulate the $CD56^{bright}CD16^{neg}$ cell subset to produce IFNγ and the production is extremely rapid and fully dependent on IL-12 released by mature DCs[8]. In vivo, murine lymph node NK cells, activated by LPS-stimulated autologous DCs migrating into the lymph node, secrete IFNγ with a peak of cytokine release after 48 h. NK cell depletion and reconstitution experiments show that NK cells provide an early source of IFNγ that is absolutely required for Th1 polarization[9]. Therefore, in a model of DC stimulation by bacterial products, NK cells play a crucial regulatory role during DC-dependent T cell priming and subsequent polarization in the T cell areas of secondary lymphoid organs.

In conclusion, early activation of NK cells by DCs activated by bacteria may play an immunoregulatory role in shaping the emerging adaptive immune responses. The ability to edit APC as well as secrete immunomodulatory cytokines might result in increased and predominantly Th1 polarized immune responses.

In summary, immunomodulatory as well as cytotoxic NK cells can be activated via dendritic cells infected by bacteria early during the immune responses. Possible interaction sites for this encounter are sites of inflammation and secondary lymphoid tissues (Figure 6.1). DC-activated NK cells might then exert effector functions against infected cells, but mainly influence the emerging adaptive immune response via APC editing and immunomodulatory cytokine secretion.

The complex cross talk occurring between these two major players of the innate immunity provides a novel mechanism by which NK cells could cooperate in the defense against bacteria.

REFERENCES

1. Trinchieri, G. (1989). Biology of natural killer cells. *Adv. Immunol.* **7**, 176–87.

2. Mingari, M. C., A. Poggi, R. Biassoni, R. Bellomo, E. Ciccone, N. Pella, *et al.* (1991). In vitro proliferation and cloning of $CD3^-CD16^+$ cells from human thymocyte precursors. *J. Exp. Med.* **174**, 21–6.

3. Rodewald, H. R., P. Moingeon, J. L. Lucich, C. Dosiou, P. Lopez, and E. L. Reinherz (1992). A population of early fetal thymocytes expressing FcγRII/III contains precursors of T lymphocytes and natural killer cells. *Cell* **69**, 139–49.

4. Lanier, L. L., H. Spits, and J. H. Phillips (1992). The developmental relationship between NK cells and T cells. *Immunol. Today* **13**, 392–9.

5. Boehm, U., T. Klamp, M. Groot, and J.C. Howard (1997). Cellular responses to interferon-gamma. *Annu. Rev. Immunol.* **15**, 749–95.

6. Mingari, M.C., C. Vitale, C. Cantoni, R. Bellomo, M. Ponte, F. Schiavetti, *et al.* (1997). Interleukin-15-induced maturation of human natural killer cells from early thymic precursors. Selective expression of CD94/NKG2A as the only HLA-class I specific inhibitory receptor. *Eur. J. Immunol.* **27**, 1374–80.

7. Ljunggren, H.G. and K. Kärre (1990). In search of the "missing self". MHC molecules and NK cell recognition. *Immunol. Today* **11**, 237–44.

8. Ferlazzo, G., M. Pack, D. Thomas, C. Paludan, D. Schmid, T. Strowig, G. Bougras, W.A. Muller, L. Moretta, and C. Munz (2004). Distinct roles of IL-12 and IL-15 in human natural killer cell activation by dendritic cells from secondary lymphoid organs. *Proc. Natl Acad. Sci. U S A* **101**, 16606–11.

9. Martin-Fontecha, A., L.L. Thomsen, S. Brett, C. Gerard, M. Lipp, A., Lanzavecchia, and F. Sallusto (2004). Induced recruitment of NK cells to lymph nodes provides IFN-gamma for T(H)1 priming. *Nat. Immunol.* **5**, 1260–5.

10. Yokoyama, W.M. and W.E. Seaman (1993). The Ly49 and NKR-P1 gene families encoding lectin-like receptors on natural killer cells: the NK gene complex. *Annu. Rev. Immunol.* **11**, 613–35.

11. Moretta, A., C. Bottino, M. Vitale, *et al.* (1996). Receptors for HLA-class I-molecules in human natural killer cells. *Annu. Rev. Immunol.* **14**, 619–48.

12. Lanier L.L. (1998). NK cell receptors. *Annu. Rev. Immunol.* **16**, 359–93.

13. Long, E.O. (1999). Regulation of immune response through inhibitory receptors. *Annu. Rev. Immunol.* **17**, 875–904.

14. Lopez-Botet, M., M. Pérez Villar, M. Carretero, *et al.* (1997). Structure and function of the CD94 C-type lectin receptor complex involved in the recognition of HLA class I molecules. *Immunol. Rev.* **155**, 165–74.

15. Khakoo, S.I., R. Rajalingam, B.P. Shum, *et al.* (2000). Rapid evolution of NK cell receptor systems demonstrated by comparison of chimpanzees and humans. *Immunity* **12**, 687–98.

16. Robertson, M.J. and J. Ritz (1990). Biology and clinical relevance of human natural killer cells. *Blood* **76**, 2421–38.

17. Campbell, J.J., S. Qin, D. Unutmaz, D. Soler, K.E. Murphy, M.R. Hodge, L. Wu, and E.C. Butcher (2001). Unique subpopulations of CD56[+] NK and NK-T peripheral blood lymphocytes identified by chemokine receptor expression repertoire. *J. Immunol.* **166**, 6477–82.

18. Jacobs, R., G. Hintzen, A. Kemper, K. Beul, S. Kempf, G. Behrens, K. W. Sykora, and R. E. Schmidt (2001). CD56^{bright} cells differ in their KIR repertoire and cytotoxic features from CD56^{dim} NK cells. *Eur. J. Immunol.* **31**, 3121–7.

19. Cooper, M. A., T. A. Fehniger, S. C. Turner, K. S. Chen, B. A. Ghaheri, T. Ghayur, W. E. Carson, and M. A. Caligiuri (2001). Human natural killer cells: a unique innate immunoregulatory role for the CD56^{bright} subset. *Blood* **97**, 3146–51.

20. Moretta, A., C. Bottino, M. Vitale, D. Pende, C. Cantoni, M. C. Mingari, R. Biassoni, and L. Moretta (2001). Activating receptors and coreceptors involved in human natural killer cell-mediated cytolysis. *Annu. Rev. Immunol.* **19**, 197–223.

21. Moretta, L., C. Bottino, D. Pende, M. C. Mingari, R. Biassoni, and A. Moretta (2002). Human natural killer cells: their origin, receptors and function. *Eur. J. Immunol.* **32**, 1205–11.

22. Moretta, L., G. Ferlazzo, M. C. Mingari, G. Melioli, and A. Moretta (2003). Human natural killer cell function and their interactions with dendritic cells. *Vaccine* **21**, Suppl. 2, S*38*.

23. Bauer, S., V. Groh, J. Wu, A. Steinle, J. H. Phillips, L. L. Lanier, and T. Spies (1999). Activation of NK cells and T cells by NKG2D, a receptor for stress- inducible MICA. *Science* **285**, 727–9.

24. Cosman, D., J. Mullberg, C. L. Sutherland, W. Chin, R. Armitage, W. Fanslow, M. Kubin, and N. J. Chalupny (2001). ULBPs, novel MHC class I-related molecules, bind to CMV glycoprotein UL16 and stimulate NK cytotoxicity through the NKG2D receptor. *Immunity* **14**, 123–33.

25. Fehniger, T. A., M. A. Cooper, G. J. Nuovo, M. Cella, F. Facchetti, M. Colonna, and M. A. Caligiuri (2003). CD56^{bright} natural killer cells are present in human lymph nodes and are activated by T cell-derived IL-2: a potential new link between adaptive and innate immunity. *Blood* **101**, 3052–7.

26. Ferlazzo, G., S. L. Lin, K. Goodman, D. Thomas, B. Morandi, W. A. Muller, A. Moretta, and C. Münz (2004). The abundant NK cells in human lymphoid tissues require activation to become cytolytic. *J. Immunol.* **172**, 1455–62.

27. Westermann, J. and R. Pabst (1992). Distribution of lymphocyte subsets and natural killer cells in the human body. *Clin. Investig.* **70**, 539–44.

28. Trepel, F. (1974). Number and distribution of lymphocytes in man. A critical analysis. *Klin. Wochenschr.* **52**, 511–15.

29. Gerosa, F., B. Baldani-Guerra, C. Nisii, V. Marchesini, G. Carra, and G. Trinchieri (2002). Reciprocal activating interaction between natural killer cells and dendritic cells. *J. Exp. Med.* **195**, 327–33.

30. Gerosa, F., A. Gobbi, P. Zorzi, S. Burg, F. Briere, G. Carra, and G. Trinchieri (2005). The reciprocal interaction of NK cells with plasmacytoid or myeloid dendritic cells profoundly affects innate resistance functions. *J. Immunol.* **174**, 727–34.

31. Piccioli, D., S. Sbrana, E. Melandri, and N. M. Valiante (2002). Contact-dependent stimulation and inhibition of dendritic cells by natural killer cells. *J. Exp. Med.* **195**, 335–41.

32. Fernandez, N. C., A. Lozier, C. Flament, P. Ricciardi-Castagnoli, D. Bellet, M. Suter, M. Perricaudet, T. Tursz, E. Maraskovsky, and L. Zitvogel (1999). Dendritic cells directly trigger NK cell functions: cross-talk relevant in innate anti-tumor immune responses in vivo. *Nat. Med.* **5**, 405–11.

33. Ferlazzo, G., M. L. Tsang, L. Moretta, G. Melioli, R. M. Steinman, and C. Munz (2002). Human dendritic cells activate resting natural killer (NK) cells and are recognized via the NKp30 receptor by activated NK cells. *J. Exp. Med.* **195**, 343–51.

34. Andrews, D. M., A. A. Scalzo, W. M. Yokoyama, M. J. Smyth, and M. A. Degli-Esposti (2003). Functional interactions between dendritic cells and NK cells during viral infection. *Nat. Immunol.* **4**, 175–81.

35. Dalod, M., T. Hamilton, R. Salomon, T. P. Salazar-Mather, S. C. Henry, J. D. Hamilton, and C. A. Biron (2003). Dendritic cell responses to early murine cytomegalovirus infection: subset functional specialization and differential regulation by interferon alpha/beta. *J. Exp. Med.* **197**, 885–98.

36. Orange, J. S. and C. A. Biron (1996). An absolute and restricted requirement for IL-12 in natural killer cell IFN-γ production and antiviral defense. Studies of natural killer and T cell responses in contrasting viral infections. *J. Immunol.* **156**, 1138–42.

37. Nguyen, K. B., T. P. Salazar-Mather, M. Y. Dalod, J. B. Van Deusen, X. Q. Wei, F. Y. Liew, M. A. Caligiuri, J. E. Durbin, and C. A. Biron (2002). Coordinated and distinct roles for IFN-alpha beta, IL-12, and IL-15 regulation of NK cell responses to viral infection. *J. Immunol.* **169**, 4279–87.

38. Granucci, F., C. Vizzardelli, N. Pavelka, S. Feau, M. Persico, E. Virzi, M. Rescigno, G. Moro, and P. Ricciardi-Castagnoli (2001). Inducible IL-2 production by dendritic cells revealed by global gene expression analysis. *Nat. Immunol.* **2**, 882–8.

39. Granucci, F., S. Feau, V. Angeli, F. Trottein, and P. Ricciardi-Castagnoli (2003). Early IL-2 production by mouse dendritic cells is the result of microbial-induced priming. *J. Immunol.* **170**, 5075–81.

40. Nishioka, Y., N. Nishimura, Y. Suzuki, and S. Sone (2001). Human monocyte-derived and CD83$^+$ blood dendritic cells enhance NK cell-mediated cytotoxicity. *Eur. J. Immunol.* **31**, 2633–41.

41. Yu, Y., M. Hagihara, K. Ando, B. Gansuvd, H. Matsuzawa, T. Tsuchiya, Y. Ueda, H. Inoue, T. Hotta, and S. Kato (2001). Enhancement of human cord blood CD34$^+$ cell-derived NK cell cytotoxicity by dendritic cells. *J. Immunol.* **166**, 1590–1600.

42. Ferlazzo, G., B. Morandi, A. D'Agostino, R. Meazza, G. Melioli, A. Moretta, and L. Moretta (2003). The interaction between NK cells and dendritic cells in bacterial infections results in rapid induction of NK cell activation and in the lysis of uninfected dendritic cells. *Eur. J. Immunol.* **33**, 306–13.

43. Munz, C., T. Dao, G. Ferlazzo, M. A. de Cos, K. Goodman, and J. W. Young (2005). Mature myeloid dendritic cell subsets have distinct roles for activation and viability of circulating human natural killer cells. *Blood* **105**, 266–73.

44. Siegal, F. P., N. Kadowaki, M. Shodell, P. A. Fitzgerald-Bocarsly, K. Shah, S. Ho, S. Antonenko, and Y. J. Liu (1999). The nature of the principal type 1 interferon-producing cells in human blood. *Science* **284**, 1835–7.

45. Bandyopadhyay, S., B. Perussia, G. Trinchieri, D. S. Miller, and S. E. Starr (1986). Requirement for HLA-DR$^+$ accessory cells in natural killing of cytomegalovirus-infected fibroblasts. *J. Exp. Med.* **164**, 180–95.

46. Dalod, M., T. P. Salazar-Mather, L. Malmgaard, C. Lewis, C. Asselin-Paturel, F. Briere, G. Trinchieri, and C. A. Biron (2002). Interferon α/β and interleukin 12 responses to viral infections: pathways regulating dendritic cell cytokine expression in vivo. *J. Exp. Med.* **195**, 517–28.

47. Feldman, M., D. Howell, and P. Fitzgerald-Bocarsly (1992). Interferon-α-dependent and -independent participation of accessory cells in natural killer cell-mediated lysis of HSV-1-infected fibroblasts. *J. Leukoc. Biol.* **52**, 473–82.

48. Diebold, S. S., M. Montoya, H. Unger, L. Alexopoulou, P. Roy, L. E. Haswell, A. Al-Shamkhani, R. Flavell, P. Borrow, and C. Reis e Sousa (2003). Viral infection switches non-plasmacytoid dendritic cells into high interferon producers. *Nature* **424**, 324–8.

49. Granucci, F., I. Zanoni, N. Pavelka, S. L. Van Dommelen, C. E. Andoniou, F. Belardelli, M. A. Degli Esposti, and P. Ricciardi-Castagnoli (2004). A contribution of mouse dendritic cell-derived IL-2 for NK cell activation. *J. Exp. Med.* **200**, 287–95.

50. Jinushi, M., T. Takehara, T. Kanto, T. Tatsumi, V. Groh, T. Spies, T. Miyagi, T. Suzuki, Y. Sasaki, and N. Hayashi (2003). Critical role of MHC class I-related chain A and B expression on IFN-α-stimulated dendritic cells in NK cell activation: impairment in chronic hepatitis C virus infection. *J. Immunol.* **170**, 1249–56.

51. Jinushi, M., T. Takehara, T. Tatsumi, T. Kanto, V. Groh, T. Spies, T. Suzuki, T. Miyagi, and N. Hayashi (2003). Autocrine/paracrine IL-15 that is required for type I IFN-mediated dendritic cell expression of MHC class I-related chain A and B is impaired in hepatitis C virus infection. *J. Immunol.* **171**, 5423–9.

52. Inaba, K., M. Inaba, M. Naito, and R. M. Steinman (1993). Dendritic cell progenitors phagocytose particulates, including bacillus Calmette-Guerin organisms, and sensitize mice to mycobacterial antigens in vivo. *J. Exp. Med.* **178**, 479–88.

53. Demangel, C., A. G. Bean, E. Martin, C. G. Feng, A. T. Kamath, and W. J. Britton (1999). Protection against aerosol *Mycobacterium tuberculosis* infection using *Mycobacterium bovis* Bacillus Calmette Guerin-infected dendritic cells. *Eur. J. Immunol.* **29**, 1972–9.

54. Cooper, M. A., T. A. Fehniger, and M. A. Caligiuri (2001). The biology of human natural killer-cell subsets. *Trends Immunol.* **22**, 633–40.

55. Mattei, F., G. Schiavoni, F. Belardelli, and D. F. Tough (2001). IL-15 is expressed by dendritic cells in response to type I IFN, double-stranded RNA, or lipopolysaccharide and promotes dendritic cell activation. *J. Immunol.* **167**, 1179–87.

56. Wilson, J. L., L. C. Heffler, J. Charo, A. Scheynius, M. T. Bejarano, and H. G. Ljunggren (1999). Targeting of human dendritic cells by autologous NK cells. *J. Immunol.* **163**, 6365–70.

57. Ferlazzo, G., C. Semino, and G. Melioli (2001). HLA class I molecule expression is up-regulated during maturation of dendritic cells, protecting them from natural killer cell-mediated lysis. *Immunol. Lett.* **76**, 37–41.

58. Moretta, A. (2002). Natural killer and dendritic cells: rendezvous in abused tissues. *Nat. Rev. Immunol.* **2**, 1–8.

59. Buentke, E., L. C. Heffler, J. L. Wilson, R. P. Wallin, C. Lofman, B. J. Chambers, H. G. Ljunggren, and A. Scheynius (2002). Natural killer and dendritic cell contact in lesional atopic dermatitis skin-Malassezia-influenced cell interaction. *J. Invest. Dermatol.* **119**, 850–7.

60. Bazan, J. F., K. B. Bacon, G. Hardiman, W. Wang, K. Soo, D. Rossi, D. R. Greaves, A. Zlotnik, and T. J. Schall (1997). A new class of membrane-bound chemokine with a CX3C motif. *Nature* **385**, 640–4.

61. Hoffmann, E., O. Dittrich-Breiholz, H. Holtmann, and M. Kracht (2002). Multiple control of interleukin-8 gene expression. *J. Leukoc. Biol.* **72**, 847–55.

62. Nishimura, M., H. Umehara, T. Nakayama, O. Yoneda, K. Hieshima, M. Kakizaki, N. Dohmae, O. Yoshie, and T. Imai (2002). Dual functions of fractalkine/CX3C ligand 1 in trafficking of perforin$^+$/granzyme B$^+$ cytotoxic effector lymphocytes that are defined by CX3CR1 expression. *J. Immunol.* **168**, 6173–80.

63. Dowdell, K.C., D.J. Cua, E. Kirkman, and S.A. Stohlman (2003). NK cells regulate CD4 responses prior to antigen encounter. *J. Immunol.* **171**, 234–9.

64. Coudert, J.D., C. Coureau, and J.C. Guery (2002). Preventing NK cell activation by donor dendritic cells enhances allospecific CD4 T cell priming and promotes Th type 2 responses to transplantation antigens. *J. Immunol.* **169**, 2979–87.

65. Ferlazzo, G. and C. Munz (2004). NK cell compartments and their activation by dendritic cells. *J. Immunol.* **172**, 1333–9.

PART III Dendritic cells and adaptive immune responses to bacteria

Peculiar ability of dendritic cells to process and present antigens from vacuolar pathogens: a lesson from *Legionella*

Sunny Shin, Catarina Nogueira and Craig R. Roy
Yale University School of Medicine

7.1 *L. PNEUMOPHILA* AND LEGIONNAIRES' DISEASE

Legionella pneumophila is a Gram-negative facultative intracellular pathogen capable of growing in both protozoan and mammalian host cells. *L. pneumophila* is found in natural and artificial water reservoirs and less often in soil and organic matter (Fields, 1996; Szymanska *et al.*, 2004). Optimal proliferation conditions for *Legionella* are those in which water temperatures are between 25°C and 42°C, calcium and magnesium salt-containing sediments are present, and are further enhanced by the presence of algae and protozoa (Szymanska *et al.*, 2004). In hostile conditions, *Legionella* and other organisms become attached to surfaces in an aquatic environment, forming a biofilm (Langmark *et al.*, 2005). *L. pneumophila* can be isolated from such natural water sources as lakes, ponds and streams; however, artificial reservoirs such as plumbing fixtures, hot water tanks, whirlpool spas and cooling towers, all possess excellent conditions for *Legionella* proliferation inside protozoan hosts and are the source of most outbreaks (Fliermans *et al.*, 1981; Yee and Wadowsky, 1982).

The first recognized outbreak of *L. pneumophila* occurred in Philadelphia in 1976 during a state convention of the American Legion (Fraser *et al.*, 1977). During this outbreak a total of 221 people contracted the disease, 34 of whom subsequently died. A new Gram-negative bacterium was isolated from both patients and the air-conditioning system of the hotel that was the source of the outbreak (McDade *et al.*, 1977). This isolated organism was named *Legionella pneumophila* (Brenner *et al.*, 1979). There are 48 different species of *Legionella* found in nature. Interestingly, *Legionella pneumophila* serogroup 1, the strain isolated in the Philadelphia outbreak, is responsible for most community-acquired outbreaks of Legionnaires' disease, followed by serogroups 4 and 6.

Legionella is acquired by inhalation of contaminated aerosolized water droplets, resulting in bacterial entry into the human lung. Bacteria that escape clearance by the action of lung cilia are ingested by alveolar macrophages. Instead of being killed by alveolar macrophages, Legionella establishes a vacuole that supports intracellular growth similar to that observed in protozoan phagocytes (Horwitz, 1983). Infection with L. pneumophila causes legionellosis, which comprises three possible distinct clinical outcomes: (1) an atypical severe pneumonia described as Legionnaires' disease; (2) a flu-like form having a mild course known as Pontiac fever; (3) an extra-pulmonary form in immunosuppressed patients, often taking a severe clinical course, with septic syndrome, coagulation disorders, acute cardiovascular deficiency and nephritis (Szymanska et al., 2004). People of any age can acquire Legionnaires' disease but the illness more often affects middle-aged and older patients, particularly those who smoke cigarettes, and individuals who have chronic respiratory diseases or renal deficiency. Also at increased risk are individuals immunocompromised by diseases such as cancer, kidney failure requiring dialysis, or diabetes. As the use of devices capable of aerosolizing contaminated water have become more common, epidemics of pneumonia that involve Legionella have also become more prevalent. Those that are exposed to water aerosols (workers of cooling towers, turbine operators and others) are at higher risk (Szymanska et al., 2004).

7.2 *LEGIONELLA* REPLICATES IN A SPECIALIZED VACUOLE DERIVED FROM THE HOST ENDOPLASMIC RETICULUM

After internalization by phagocytes, L. pneumophila evades transport to a lysosomal compartment, and establishes a replicative niche distinct from the normal endocytic pathway (Roy and Tilney, 2002). The replicative vacuole established by L. pneumophila is remodeled by early secretory vesicles traveling between the host endoplasmic reticulum (ER) and Golgi apparatus. The mature vacuole in which L. pneumophila replicates is decorated with ribosomes and is similar to the host ER (Tilney et al., 2001). The ability of L. pneumophila to alter trafficking of the vacuole in which it resides is dependent upon a bacterial secretion apparatus called the Dot/Icm system (Segal et al., 1998; Vogel et al., 1998). Upon encountering host cells, the Dot/Icm system transfers bacterial proteins directly into the host cytosol (Nagai et al., 2002; Chen et al., 2004; Luo and Isberg, 2004). These bacterial proteins engage host factors involved in vesicular transport and redirect the vacuole containing *L. pneumophila* to prevent endocytic maturation and promote remodeling by ER-derived vesicles (Horwitz, 1983; Tilney *et al.*, 2001; Kagan and Roy, 2002). Intracellular growth mutants of *L. pneumophila* that are defective in the Dot/Icm system are unable to alter transport of their vacuole and reside in vacuoles that mature similarly to phagosomes containing inert particles or non-pathogenic bacteria (Horwitz, 1987; Marra *et al.*, 1992).

7.3 THE BIRC1E PROTEIN MEDIATES HOST RESISTANCE TO *LEGIONELLA*

Primary macrophages isolated from humans, guinea pigs, and hamsters permit the intracellular replication of *L. pneumophila*, whereas, macrophages isolated from most mouse strains do not allow for *L. pneumophila* growth. Interestingly, macrophages from A/J mice are permissive for *L. pneumophila* growth (Yamamoto *et al.*, 1988). This phenotype translates to the in vivo setting, with A/J mice capable of supporting *L. pneumophila* infection in a manner comparable to the human course of disease (Brieland *et al.*, 1994). The susceptibility of A/J mice to *L. pneumophila* infection is controlled by a single genetic locus called *Lgn1*. Birc1e (also called Naip5) was positionally cloned as the gene within this locus responsible for the permissive phenotype (Diez *et al.*, 2003; Wright *et al.*, 2003). Genetic polymorphisms in Birc1e thus appear to control the ability of *L. pneumophila* to replicate in mouse macrophages (Diez *et al.*, 2003; Wright *et al.*, 2003). Birc1e is a member of the recently discovered Nod-LRR (nucleotide oligomerization domain-leucine-rich repeat) family of cytosolic pathogen recognition receptors (PRRs) (Inohara *et al.*, 2004; Ting and Williams, 2005). Nod-LRRs are thought to be responsible for dealing with intracellular pathogens that might evade detection by Toll-like receptors (TLRs). The Nod proteins have some homology to TLRs in that they contain LRR domains thought to be responsible for detection of bacterial products (Inohara *et al.*, 2004; Ting and Williams, 2005). Birc1e is predicted to contain three baculoviral inhibitors of apoptosis repeat (BIR) domains, a NOD region and an LRR domain (Inohara *et al.*, 2004; Ting and Williams, 2005). The presence of BIR domains makes Birc1e similar to proteins in the inhibitors of apoptosis (IAP) family. The molecular mechanism by which Birc1e controls host resistance to *L. pneumophila* is currently unknown. A detailed analysis of the fate of *L. pneumophila* in infected C57Bl/6 macrophages, which contain the restrictive *Lgn1* allele, found that there appears to be multifactorial resistance to *L. pneumophila*. This resistance is mediated in part through retrafficking of the *L. pneumophila* phagosome to the lysosome and the induction of apoptosis in infected cells (Derre and Isberg, 2004).

7.4 INNATE IMMUNE RESPONSES TO *LEGIONELLA* BY TOLL-LIKE RECEPTORS

Work examining the ability of *L. pneumophila* lipopolysaccharide (LPS) to activate TLR signaling in vitro indicates that TLR2 rather than TLR4 is primarily responsible for initiating cellular responses to *L. pneumophila* LPS (Girard *et al.*, 2003). Highly purified LPS is able to activate TLR2 signaling in macrophages, while TLR4-deficient macrophages show defects in their ability to respond to *L. pneumophila* LPS (Girard *et al.*, 2003). This result is supported by data showing that intracellular growth of *L. pneumophila* is enhanced within TLR2$^{-/-}$ macrophages compared to WT and TLR4$^{-/-}$ macrophages (Akamine *et al.*, 2005). This increase in intracellular growth is accompanied by a significant decrease in the production of IL-12, which is essential for clearing *L. pneumophila* infection (Akamine *et al.*, 2005). These studies indicate that TLR2, rather than TLR4, plays a principal role in initiating immune responses to *L. pneumophila*. However, the importance of TLR2 in controlling *Legionella* replication in vivo is not yet established.

TLR5 has also been implicated in the innate immune response to *L. pneumophila*. TLR5 is responsible for sensing bacterial flagellin (Smith *et al.*, 2003). *L. pneumophila* flagellin activates TLR5 signaling pathways, resulting in the induction of IL-8 production by lung epithelial cell lines (Hawn *et al.*, 2003). Additionally, epidemiological studies provide a link between a common human genetic polymorphism, which produces a truncated TLR5 protein that cannot respond to flagellin, and susceptibility to *L. pneumophila*-induced pneumonia in humans (Hawn *et al.*, 2003). Whether TLR5 is required for the initiation of the immune response to *L. pneumophila* is unclear. It would be expected that other TLRs, such as TLR9, which recognizes bacterial CpG DNA, would also be involved. Although the role of the TLRs in initiating the immune response to *L. pneumophila* infection has not yet been addressed, it is expected that some of the TLRs would be involved in the initiation of cytokine production.

7.5 INDUCTION OF CYTOKINES FOLLOWING
LEGIONELLA INFECTION

Humans infected with *L. pneumophila* develop a predominantly T1 cytokine profile, which mimics the Th1 cytokine response (Tateda *et al.*, 1998). The T1/T2 cytokine balance in infected hosts appears to be critical for the outcome of *L. pneumophila* infection. In mice, neutrophils have been shown to play a prominent role in determining subsequent T1/T2 host responses to *L. pneumophila* (Tateda *et al.*, 2001). Depletion of neutrophils results in a hundred-fold increase in susceptibility of mice to lethal pneumonia. This effect was not due to direct neutrophil killing of *L. pneumophila*. Instead, neutrophils appeared to modulate the cytokine profile of infected mice. These mice had a significant decrease in IFN-γ and IL-12 production and an increase in the T2 cytokines IL-4 and IL-10 (Tateda *et al.*, 2001). Additionally, IL-10 treatment can enhance the growth of *L. pneumophila* in human monocytes and reverse the protective effect of IFN-γ (Park and Skerrett, 1996). This indicates that development of a T1 response is required for clearance of *L. pneumophila* infection.

IFN-γ is a primary T1 cytokine and is critical in the activation of macrophages. It is required for the resolution of infections caused by a variety of bacteria, such as *Salmonella typhimurium* and *Mycobacterium tuberculosis*. IFN-γ is also critical for the clearance of *L. pneumophila* infection in mice. Mice depleted of IFN-γ are unable to clear *L. pneumophila* infection (Brieland *et al.*, 1994). This is supported by experiments demonstrating that mice genetically deficient in IFN-γ are also unable to clear infection (Heath *et al.*, 1996; Shinozawa *et al.*, 2002). Instead, they develop persistent and replicative *L. pneumophila* infections, which disseminate to the spleen (Heath *et al.*, 1996). In vivo, natural killer cells appear to be the primary source of IFN-γ production during *L. pneumophila* infection (Blanchard *et al.*, 1988; Deng *et al.*, 2001). IFN-γ is also induced in non-permissive macrophages following in vitro *L. pneumophila* infection (Salins *et al.*, 2001).

7.6 ADAPTIVE IMMUNE RESPONSES TO *LEGIONELLA* INFECTION

Studies examining the immune response to *L. pneumophila* have revealed that adaptive immunity develops during the course of infection. Humans infected with *L. pneumophila* often produce a specific humoral response. Infected mice also develop a humoral response to *L. pneumophila*.

Infection of A/J mice results in the recruitment of B lymphocytes into the lung and the production of anti-*L. pneumophila* antibodies (Brieland *et al.*, 1996). Additionally, guinea pigs sublethally infected with *L. pneumophila* also develop a robust humoral response (Breiman and Horwitz, 1987). However, the precise role played by the antibody response in clearing infection or in conferring a protective immune response is currently unclear. In fact, it has been reported that antibodies may actually promote *L. pneumophila* infection by increasing uptake of *L. pneumophila* in macrophages through opsonization (Horwitz and Silverstein, 1981).

Several studies indicate that *L. pneumophila*-infected animals develop an acquired Th1 immune response, which appears to result in the clearance of infection and in the development of a memory T cell response. Splenic lymphocytes from *L. pneumophila*-infected guinea pigs proliferate robustly in response to *L. pneumophila* antigens in vitro (Breiman and Horwitz, 1987). Guinea pigs also develop delayed-type hypersensitivity (DTH), a classic sign of Th1 immunity, which results in an inflammatory reaction at the site of intercutaneous infection of *L. pneumophila* (Weeratna *et al.*, 1994). The role of T cells during a primary *L. pneumophila* infection has been examined by depletion of either CD4, CD8 or both T cell subsets (Susa *et al.*, 1998). Mice depleted of CD4 or CD8 T cells show a slight delay in clearance of *L. pneumophila*, whereas mice depleted of both CD4 and CD8 T cells are significantly impaired in their ability to control *L. pneumophila* infection, resulting in increased lethality (Susa *et al.*, 1998). This indicates a critical role for both CD4 and CD8 T cells in control of *L. pneumophila* infection. However, it is unclear whether this primary T cell response is directed mainly against *L. pneumophila* antigens or whether it is due to an intrinsic requirement for T cell cytokine production.

7.7 RESTRICTION OF *LEGIONELLA* GROWTH IN ACTIVATED MACROPHAGES

It has long been known that treating macrophages with IFN-γ enhances their ability to inhibit the growth of many intracellular bacterial pathogens. Similarly, IFN-γ also restricts *L. pneumophila* growth in macrophages (Bhardwaj *et al.*, 1986; Nash *et al.*, 1988). In the case of *L. pneumophila*, it is thought that intracellular growth restriction is mediated in part by restricting the availability of intracellular iron (Byrd and Horwitz, 1989). *L. pneumophila* growth within macrophages normally requires the presence of iron, which is transported by the transferrin receptor (Byrd and Horwitz, 1989). Macrophages activated by IFN-γ downregulate expression of transferrin receptor, which limits the amount of iron and thus inhibits *L. pneumophila* growth (Byrd and Horwitz, 1989). Addition of iron to IFN-γ treated macrophages can reverse the restriction of *L. pneumophila* replication (Byrd and Horwitz, 1991).

Macrophages activated with IFN-γ are also able to override the ability of pathogens such as *Mycobacterium avium* to alter trafficking of their phagosomes, resulting in phagolysosome fusion (Schaible *et al.*, 1998). This appears to be the case for *L. pneumophila* as well. In IFN-γ activated macrophages, the *L. pneumophila* phagosome appears to mature into a phagolysosome, as evidenced by the acquisition of the lysosomal markers LAMP-2 and cathepsin D (Santic *et al.*, 2005). The mechanism of how IFN-γ is able to initiate the retrafficking of the *L. pneumophila* phagosome is unknown.

In addition, the ability of IFN-γ-treated macrophages to inhibit *L. pneumophila* growth may also be mediated by the production of reactive nitric oxide or reactive oxygen intermediates, both of which are essential anti-microbial effectors of macrophages. However, the importance of nitric oxide and oxygen intermediates in controlling *L. pneumophila* infection is unclear. Treatment of activated macrophages with superoxide dismutase and catalase does not affect the inhibition of *L. pneumophila* growth, suggesting that reactive oxygen intermediates are dispensable for controlling *Legionella* replication (Klein *et al.*, 1991). Treatment of A/J mice with the NO synthetase inhibitor N-monomethyl L-arginine (L-MMA) inhibits their ability to clear *L. pneumophila* infection (Brieland *et al.*, 1995). However, treatment of BALB/c macrophages or mice with L-MMA does not significantly inhibit clearance of *L. pneumophila* infection (Heath *et al.*, 1996; Yamamoto *et al.*, 1996). Using C57BL/6 mice genetically deficient for NADPH oxidase, it was found that NADPH oxidase appears to be important in early control of *L. pneumophila* infection, but is dispensable at later time points (Saito *et al.*, 2001). Collectively, these data indicate that the role of nitric oxide intermediates in controlling *L. pneumophila* growth may differ between susceptible and resistant macrophages. These results also indicate that mechanisms other than the nitric oxide burst are also responsible for controlling *L. pneumophila* replication. It is possible that reactive oxygen and nitric oxide intermediates may have redundant roles in inhibiting *L. pneumophila* growth. This possibility has not yet been formally tested.

In addition to IFN-γ, type I IFN (IFN-αβ) treatment of permissive macrophages results in an inhibition of intracellular *L. pneumophila* replication (Schiavoni *et al.*, 2004). Depletion of IFN-αβ from non-permissive macrophage cultures enhances their susceptibility to *L. pneumophila* infection, indicating a role for endogenous IFN-αβ in restricting intra-cellular *L. pneumophila* growth (Schiavoni *et al.*, 2004). IFN-αβ appears to act independently of IFN-γ (Schiavoni *et al.*, 2004). The molecular mechanisms underlying how IFN-αβ controls intracellular *L. pneumophila* replication and whether it is through a mechanism similar to that used by IFN-γ is unknown.

7.8 *LEGIONELLA* GROWTH RESTRICTION BY MURINE DENDRITIC CELLS

L. pneumophila has been used as a model organism to dissect the mechanisms by which immune responses are generated against vacuolar pathogens. To understand how immunity against vacuolar pathogens is generated, it is important to determine how T cell responses are primed and the mechanisms by which bacterial antigens become available for presentation on MHC class II. Although the interactions between *L. pneumophila* and macrophages have been extensively studied over the past years, the mechanisms underlying the development of adaptive immunity to this organism remain elusive (Kikuchi *et al.*, 2004). Recent studies have revealed that DCs are also likely to play an important role in the initiation of the adaptive immune response to *L. pneumophila* (Neild and Roy, 2003).

Using DC derived from the permissive A/J mouse strain, it was determined that the intracellular behavior of *L. pneumophila* inside dendritic cells (DCs) differs from that in macrophages. Specifically, in contrast to macrophages, DCs inhibit replication of *Legionella*. Twelve hours after infecting DCs with *L. pneumophila*, the number of vacuoles harboring replicating *L. pneumophila*, as determined by the vacuole having more than two bacterial cells, is extremely low. By contrast, when the same experiment is performed in macrophages, roughly 75 per cent of all vacuoles containing *L. pneumophila* harbor more than four bacteria due to robust intracellular growth (Neild and Roy, 2004). No significant host cell killing was observed after infection of DCs by wild type *L. pneumophila*, indicating that the DCs restrict the growth of *L. pneumophila* by a mechanism independent of cell death (Neild and Roy, 2004).

To better understand how DCs restrict the growth of *L. pneumophila*, morphological studies were conducted to determine whether vacuoles containing bacteria traffic differently in DCs compared to macrophages. These studies showed that vacuoles containing *L. pneumophila* traffic to the ER in both DCs and macrophages, forming similar ER-derived vacuoles. ER vesicles, ribosomes and mitochondria are associated with the membrane of the vacuole established by *L. pneumophila*, in contrast to the naked membrane vacuole containing *dotA* strains which fuses with the lysosome. Inside DCs, *L. pneumophila* cannot replicate but it remains metabolically active (Neild and Roy, 2004). Thus, the mechanism by which DCs restrict growth of *L. pneumophila* occurs after establishment of the ER vacuole and after *L. pneumophila* begins to respond to this new environment.

7.9 PRESENTATION OF *LEGIONELLA* ANTIGENS BY INFECTED DENDRITIC CELLS

Although *L. pneumophila* reside within an ER-derived compartment in DCs, bacterial antigens are still presented on MHC class II molecules. DCs infected with *L. pneumophila* stimulate immune CD4[+] T cells, indicating that *L. pneumophila* antigens are being processed and presented on MHC class II (Neild and Roy, 2003). Interestingly, in vivo studies show that after immunization of mice with wild type or *dotA* mutant *L. pneumophila*, two sets of T cell responses are generated: those that can respond to common determinants presented by both wild-type and *dotA L. pneumophila* and another subset of T cells that seem specific for bacterial determinants synthesized and presented after wild-type *L. pneumophila* establish an ER-derived vacuole (Neild and Roy, 2004). These data reveal that DCs have the ability to restrict the intracellular growth of *L. pneumophila* and present bacteria antigens on MHC II, which would be expected to be an important property for priming effective T-cell mediated responses to vacuolar pathogens.

In a second study, it was shown that DCs pulsed with dead *L. pneumophila*, but not with live bacteria, undergo maturation, resulting in the upregulation of MHC II, costimulatory, adhesion and signaling molecules such as CD40, CD54 (ICAM-1), CD80 (B7-1) and CD86 (B7-2) (Kikuchi *et al.*, 2004). Apart from IL-1β, the same event is also true for proinflammatory cytokine expression, including IL-12, TNF-α and IL-6. Moreover, when adoptively transferred, dead *L. pneumophila*-pulsed DCs induce *L. pneumophila*-specific protective immunity in a manner dependent on MHC II Ag presentation to CD4[+] T cells (Kikuchi *et al.*, 2004).

These findings provide a better understanding of the role of DCs and the critical link between the innate and adaptive immune responses to vacuolar pathogens. The ability of phagocytes to restrict *L. pneumophila* growth appears to be the key to host immunity (Figure 7.1). However, the mechanisms by which DCs restrict *L. pneumophila* intracellular growth are unknown. A number of studies show that control of iron availability is required for restricting *L. pneumophila* growth in activated macrophages but it still remains unclear which host gene products are involved in restriction of intracellular *L. pneumophila* growth in both macrophages and DCs (Byrd and Horwitz, 1989; Neild and Roy, 2004). It is also unknown whether the ability of DCs to restrict bacterial replication depends upon the upregulation of DC effector functions, or whether the restriction of *L. pneumophila* growth by DCs occurs independently of DC maturation. Thus, it is currently of great interest to identify the potential genes involved in restricting *L. pneumophila* replication within DCs.

7.10 FUTURE QUESTIONS IN *LEGIONELLA* IMMUNITY

It is becoming increasingly clear that DCs and macrophages have mechanisms for sensing *L. pneumophila* products in the cytosol and that detection of these products can limit growth of the bacterium. During the establishment of the *L. pneumophila* vacuole, the *L. pneumophila* Dot/Icm type IV secretion system actively transports bacterial products into the host cytosol (Nagai *et al.*, 2002; Chen *et al.*, 2004; Luo and Isberg, 2004). Some of these bacterial components are most likely sensed by members of the Nod-LRR protein family. In particular, the Nod-LRR family member Birc1e, which has been mapped as a genetic susceptibility locus for *L. pneumophila* infection, is thought to be involved in the detection of *L. pneumophila* infection (Diez *et al.*, 2003; Wright *et al.*, 2003). Birc1e has been shown to be involved in restricting the growth of *L. pneumophila* inside macrophages (Diez *et al.*, 2003; Wright *et al.*, 2003; Derre and Isberg, 2004). How Birc1e accomplishes this is unclear, but it is expected that Birc1e regulates an intracellular effector pathway which controls *L. pneumophila* replication and trafficking. The role of other Nod-LRR proteins in the immune response to *L. pneumophila* has yet to be determined. It is possible that Birc1e or other Nod-LRR proteins are involved in the restriction of *L. pneumophila* growth in DCs as well.

In addition to Birc1e, other host factors such as IFN-γ activation, iron sequestration and reactive oxygen and nitric oxide intermediates have been proposed to play a role in limiting *L. pneumophila* replication in macrophages. Further work needs to be done on whether these factors

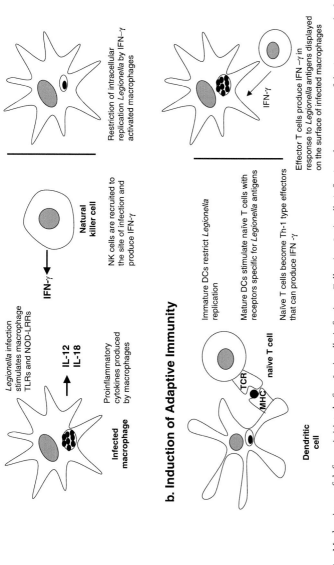

a. Clearance by the Innate Immune System

Infected macrophage

Legionella infection stimulates macrophage TLRs and NOD-LRRs

IL-12
IL-18

Proinflammatory cytokines produced by macrophages

IFN-γ

Natural killer cell

NK cells are recruited to the site of infection and produce IFN-γ

Restriction of intracellular replication *Legionella* by IFN-γ activated macrophages

b. Induction of Adaptive Immunity

Dendritic cell

TCR
MHC
naïve T cell

Immature DCs restrict *Legionella* replication

Mature DCs stimulate naïve T cells with receptors specific for *Legionella* antigens

Naïve T cells become Th-1 type effectors that can produce IFN-γ

IFN-γ

Effector T cells produce IFN-γ in response to *Legionella* antigens displayed on the surface of infected macrophages

Figure 7.1. Mechanisms of defence initiated after *Legionella* infection. Following *Legionella* infection the two arms of the innate and adaptive immune response are initiated. a. Clearance by the Innate Immune System (upper panel). *Legionella* infection stimulates macrophage TLRs and NOD-LRRs. This leads to the production of proinflammatory cytokines (IL-12 and IL-18) by macrophages. NK cells are recruited to the site of infection and produce IFN-γ. Restriction of intracellular replication *Legionella* by IFN-γ activated macrophages. b. Induction of Adaptive Immunity (lower panel). Immature DCs restrict *Legionella* replication and are induced to maturation. Mature DCs stimulate naïve T cells with receptors specific for *Legionella* antigens. Naïve T cells become Th-1 type effectors that can produce IFN-γ in response to *Legionella* antigens displayed on the surface of infected macrophages.

are also utilized by DCs to control *L. pneumophila* growth. Most likely, similar mechanisms are utilized by both DCs and activated macrophages to control *L. pneumophila* growth. A comparative proteomic analysis of the *L. pneumophila* phagosome in DCs, IFN-γ-activated macrophages and non-activated macrophages may identify novel host factors involved in the restriction of *L. pneumophila* growth. Understanding how murine macrophages and dendritic cells are able to restrict the intracellular growth of *L. pneumophila* will also provide insight into how the immune response is able to control other intracellular pathogens such as *Chlamydia trachomatis* and *Mycobacterium tuberculosis*.

Animals infected with *L. pneumophila* can develop a robust and protective T cell response. Development of a T cell response would require the presentation of *L. pneumophila* antigens on MHC class I and II molecules. How processing and presentation of *L. pneumophila* antigens on MHC molecules occurs in light of the observation that the *L. pneumophila* vacuole is distinct from the endocytic pathway is unknown. We know that DCs are able to inhibit the intracellular replication of *L. pneumophila*, yet *L. pneumophila* is not destroyed via lysosomal fusion (Neild and Roy, 2003). Although *L. pneumophila* is able to reside in a phagosome that inhibits fusion with lysosomes, DCs infected with *L. pneumophila* are capable of eliciting a *L. pneumophila*-specific CD4 T cell response (Neild and Roy, 2003). This raises the question of how *L. pneumophila* antigens are processed and presented on MHC class II molecules. Perhaps MHC class II processing and presentation stems from the *L. pneumophila* phagosome. Alternatively, *L. pneumophila* antigens may be able to leave the phagosome and enter the MHC class II presentation pathway elsewhere in the cell.

As for MHC class I presentation, it would be expected that *L. pneumophila* antigens can become accessible to the MHC class I presentation pathway via the transport of *L. pneumophila* proteins through the Dot/Icm type IV secretory apparatus into the cytosol, where they are then processed by the proteasome and transported via TAP into the ER for loading onto MHC class I molecules. Alternatively, *L. pneumophila* antigens might be processed by alternative pathways, such as the cross-presentation pathway, which would require the host cell-mediated transport of the lumenal contents of the phagosome and their processing and loading onto MHC class I. Answering these questions will most likely require the identification of the *L. pneumophila* antigens that are being presented. Studying how *L. pneumophila* antigens become accessible to the MHC class I and II pathways most likely will reveal novel antigen processing and presentation pathways. Additionally, it may provide clues into how the antigens of other vacuolar pathogens are processed as well.

The ease of genetic manipulation of *Legionella* along with the existence of known mutants with defects in their ability to modify phagosomal trafficking provides an excellent system to examine how immune responses are generated against bacteria in distinct subcellular compartments. Furthermore, the recent observations on the differential ability of macrophages and DCs to control *Legionella* growth provides investigators with the ability to dissect both the host and bacterial sides of the *Legionella*–phagocyte interaction. These studies will doubtless lead to new insights into the mechanisms of how phagocytes control replication of bacterial pathogens.

REFERENCES

Akamine, M., F. Higa *et al.* (2005). Differential roles of Toll-like receptors 2 and 4 in in vitro responses of macrophages to *Legionella pneumophila*. *Infect. Immun.* **73**(1), 352–61.

Bhardwaj, N., T. W. Nash *et al.* (1986). Interferon-gamma-activated human monocytes inhibit the intracellular multiplication of *Legionella pneumophila*. *J. Immunol.* **137**(8), 2662–9.

Blanchard, D. K., H. Friedman *et al.* (1988). Role of gamma interferon in induction of natural killer activity by *Legionella pneumophila* in vitro and in an experimental murine infection model. *Infect. Immun.* **56**(5), 1187–93.

Breiman, R. F. and M. A. Horwitz (1987). Guinea pigs sublethally infected with aerosolized *Legionella pneumophila* develop humoral and cell-mediated immune responses and are protected against lethal aerosol challenge. A model for studying host defense against lung infections caused by intracellular pathogens. *J. Exp. Med.* **165**(3), 799–811.

Brenner, D. J., A. G. Steigerwalt *et al.* (1979). Classification of the Legionnaires' disease bacterium: *Legionella pneumophila*, genus novum, species nova, of the family *Legionellaceae*, familia nova. *Ann. Intern. Med.* **90**(4), 656–8.

Brieland, J., P. Freeman *et al.* (1994). Replicative *Legionella pneumophila* lung infection in intratracheally inoculated A/J mice. A murine model of human Legionnaires' disease. *Am. J. Pathol.* **145**(6), 1537–46.

Brieland, J. K., L. A. Heath *et al.* (1996). Humoral immunity and regulation of intrapulmonary growth of *Legionella pneumophila* in the immunocompetent host. *J. Immunol.* **157**(11), 5002–8.

Brieland, J. K., D. G. Remick *et al.* (1995). In vivo regulation of replicative *Legionella pneumophila* lung infection by endogenous tumor necrosis factor alpha and nitric oxide. *Infect. Immun.* **63**(9), 3253–8.

Byrd, T. F. and M. A. Horwitz (1989). Interferon gamma-activated human monocytes downregulate transferrin receptors and inhibit the intracellular multiplication of *Legionella pneumophila* by limiting the availability of iron. *J. Clin. Invest.* **83**(5), 1457–65.

Byrd, T. F. and M. A. Horwitz (1991). Lactoferrin inhibits or promotes *Legionella pneumophila* intracellular multiplication in nonactivated and interferon gamma-activated human monocytes depending upon its degree of iron saturation. Iron-lactoferrin and nonphysiologic iron chelates reverse monocyte activation against *Legionella pneumophila*. *J. Clin. Invest.* **88**(4), 1103–12.

Chen, J., K. S. de Felipe *et al.* (2004). *Legionella* effectors that promote nonlytic release from protozoa. *Science* **303**(5662), 1358–61.

Deng, J. C., K. Tateda *et al.* (2001). Transient transgenic expression of gamma interferon promotes *Legionella pneumophila* clearance in immunocompetent hosts. *Infect. Immun.* **69**(10), 6382–90.

Derre, I. and R. R. Isberg (2004). Macrophages from mice with the restrictive *Lgn1* allele exhibit multifactorial resistance to *Legionella pneumophila*. *Infect. Immun.* **72**(11), 6221–9.

Diez, E., S. H. Lee *et al.* (2003). *Birc1e* is the gene within the *Lgn1* locus associated with resistance to *Legionella pneumophila*. *Nat. Genet.* **33**(1), 55–60.

Fields, B. S. (1996). The molecular ecology of *Legionellae*. *Trends Microbiol.* **4**(7), 286–90.

Fliermans, C. B., W. B. Cherry *et al.* (1981). Ecological distribution of *Legionella pneumophila*. *Appl. Environ. Microbiol.* **41**(1), 9–16.

Fraser, D. W., T. R. Tsai *et al.* (1977). Legionnaires' disease: description of an epidemic of pneumonia. *N. Engl. J. Med.* **297**(22), 1189–97.

Girard, R., T. Pedron *et al.* (2003). Lipopolysaccharides from *Legionella* and *Rhizobium* stimulate mouse bone marrow granulocytes via Toll-like receptor 2. *J. Cell Sci.* **116**(Pt 2), 293–302.

Hawn, T. R., A. Verbon *et al.* (2003). A common dominant TLR5 stop codon polymorphism abolishes flagellin signaling and is associated with susceptibility to legionnaires' disease. *J. Exp. Med.* **198**(10), 1563–72.

Heath, L., C. Chrisp *et al.* (1996). Effector mechanisms responsible for gamma interferon-mediated host resistance to *Legionella pneumophila* lung infection: the role of endogenous nitric oxide differs in susceptible and resistant murine hosts. *Infect. Immun.* **64**(12), 5151–60.

Horwitz, M. A. (1983). Formation of a novel phagosome by the Legionnaires' disease bacterium (*Legionella pneumophila*) in human monocytes. *J. Exp. Med.* **158**(4), 1319–31.

Horwitz, M. A. (1987). Characterization of avirulent mutant *Legionella pneumophila* that survive but do not multiply within human monocytes. *J. Exp. Med.* **166**(5), 1310–28.

Horwitz, M. A. and S. C. Silverstein (1981). Interaction of the legionnaires' disease bacterium (*Legionella pneumophila*) with human phagocytes. II. Antibody promotes binding of *L. pneumophila* to monocytes but does not inhibit intracellular multiplication. *J. Exp. Med.* **153**(2), 398–406.

Inohara, N., M. Chamaillard *et al.* (2005). NOD-LRR proteins: role in host–microbial interactions and inflammatory disease. *Annu. Rev. Biochem.* **74**: 355–83.

Kagan, J. C. and C. R. Roy (2002). *Legionella* phagosomes intercept vesicular traffic from endoplasmic reticulum exit sites. *Nat. Cell Biol.* **4**(12), 945–54.

Kikuchi, T., T. Kobayashi *et al.* (2004). Dendritic cells pulsed with live and dead *Legionella pneumophila* elicit distinct immune responses. *J. Immunol.* **172**(3), 1727–34.

Klein, T. W., Y. Yamamoto *et al.* (1991). Interferon-gamma induced resistance to *Legionella pneumophila* in susceptible A/J mouse macrophages. *J. Leukoc. Biol.* **49**(1), 98–103.

Langmark, J., M. V. Storey *et al.* (2005). Accumulation and fate of microorganisms and microspheres in biofilms formed in a pilot-scale water distribution system. *Appl. Environ. Microbiol.* **71**(2), 706–12.

Luo, Z. Q. and R. R. Isberg (2004). Multiple substrates of the *Legionella pneumophila* Dot/Icm system identified by interbacterial protein transfer. *Proc. Natl Acad. Sci. U S A* **101**(3), 841–6.

Marra, A., S. J. Blander *et al.* (1992). Identification of a *Legionella pneumophila* locus required for intracellular multiplication in human macrophages. *Proc. Natl Acad. Sci. U S A* **89**(20), 9607–11.

McDade, J. E., C. C. Shepard *et al.* (1977). Legionnaires' disease: isolation of a bacterium and demonstration of its role in other respiratory disease. *N. Engl. J. Med.* **297**(22), 1197–203.

Nagai, H., J. C. Kagan *et al.* (2002). A bacterial guanine nucleotide exchange factor activates ARF on *Legionella* phagosomes. *Science* **295**(5555), 679–82.

Nash, T. W., D. M. Libby *et al.* (1988). IFN-gamma-activated human alveolar macrophages inhibit the intracellular multiplication of *Legionella pneumophila*. *J. Immunol.* **140**(11), 3978–81.

Neild, A. L. and C. R. Roy (2003). *Legionella* reveal dendritic cell functions that facilitate selection of antigens for MHC class II presentation. *Immunity* **18**(6), 813–23.

Neild, A. L. and C. R. Roy (2004). Immunity to vacuolar pathogens: what can we learn from *Legionella*? *Cell Microbiol.* **6**(11), 1011–18.

Park, D. R. and S. J. Skerrett (1996). IL-10 enhances the growth of *Legionella pneumophila* in human mononuclear phagocytes and reverses the protective effect of IFN-gamma: differential responses of blood monocytes and alveolar macrophages. *J. Immunol.* **157**(6), 2528–38.

Roy, C. R. and L. G. Tilney (2002). The road less traveled: transport of *Legionella* to the endoplasmic reticulum. *J. Cell Biol.* **158**(3), 415–19.

Saito, M., H. Kajiwara *et al.* (2001). Fate of *Legionella pneumophila* in macrophages of C57BL/6 chronic granulomatous disease mice. *Microbiol. Immunol.* **45**(7), 539–41.

Salins, S., C. Newton *et al.* (2001). Differential induction of gamma interferon in *Legionella pneumophila*-infected macrophages from BALB/c and A/J mice. *Infect. Immun.* **69**(6), 3605–10.

Santic, M., M. Molmeret *et al.* (2005). Maturation of the *Legionella pneumophila*-containing phagosome into a phagolysosome within gamma interferon-activated macrophages. *Infect. Immun.* **73**(5), 3166–71.

Schaible, U. E., S. Sturgill-Koszycki *et al.* (1998). Cytokine activation leads to acidification and increases maturation of *Mycobacterium avium*-containing phagosomes in murine macrophages. *J. Immunol.* **160**(3), 1290–6.

Schiavoni, G., C. Mauri *et al.* (2004). Type I IFN protects permissive macrophages from *Legionella pneumophila* infection through an IFN-gamma-independent pathway. *J. Immunol.* **173**(2), 1266–75.

Segal, G., M. Purcell *et al.* (1998). Host cell killing and bacterial conjugation require overlapping sets of genes within a 22-kb region of the *Legionella pneumophila* genome. *Proc. Natl Acad. Sci. U S A* **95**(4), 1669–74.

Shinozawa, Y., T. Matsumoto *et al.* (2002). Role of interferon-gamma in inflammatory responses in murine respiratory infection with *Legionella pneumophila*. *J. Med. Microbiol.* **51**(3), 225–30.

Smith, K. D., E. Andersen-Nissen *et al.* (2003). Toll-like receptor 5 recognizes a conserved site on flagellin required for protofilament formation and bacterial motility. *Nat. Immunol.* **4**(12), 1247–53.

Susa, M., B. Ticac *et al.* (1998). *Legionella pneumophila* infection in intratracheally inoculated T cell-depleted or -nondepleted A/J mice. *J. Immunol.* **160**(1), 316–21.

Szymanska, J., L. Wdowiak *et al.* (2004). Microbial quality of water in dental unit reservoirs. *Ann. Agric. Environ. Med.* **11**(2), 355–8.

Tateda, K., T. Matsumoto *et al.* (1998). Serum cytokines in patients with *Legionella* pneumonia: relative predominance of Th1-type cytokines. *Clin. Diagn. Lab. Immunol.* **5**(3), 401–3.

Tateda, K., T. A. Moore *et al.* (2001). Early recruitment of neutrophils determines subsequent T1/T2 host responses in a murine model of *Legionella pneumophila* pneumonia. *J. Immunol.* **166**(5), 3355–61.

Tilney, L. G., O. S. Harb *et al.* (2001). How the parasitic bacterium *Legionella pneumophila* modifies its phagosome and transforms it into rough ER: implications for conversion of plasma membrane to the ER membrane. *J. Cell Sci.* **114**(Pt 24), 4637–50.

Ting, J. P. and K. L. Williams (2005). The CATERPILLER family: An ancient family of immune/apoptotic proteins. *Clin. Immunol.* **115**(1), 33–7.

Vogel, J. P., H. L. Andrews *et al.* (1998). Conjugative transfer by the virulence system of *Legionella pneumophila*. *Science* **279**(5352), 873–6.

Weeratna, R., D. A. Stamler *et al.* (1994). Human and guinea pig immune responses to *Legionella pneumophila* protein antigens OmpS and Hsp60. *Infect. Immun.* **62**(8), 3454–62.

Wright, E. K., S. A. Goodart *et al.* (2003). Naip5 affects host susceptibility to the intracellular pathogen *Legionella pneumophila*. *Curr. Biol.* **13**(1), 27–36.

Yamamoto, Y., T. W. Klein *et al.* (1996). Immunoregulatory role of nitric oxide in *Legionella pneumophila*-infected macrophages. *Cell. Immunol.* **171**(2), 231–9.

Yamamoto, Y., T. W. Klein *et al.* (1988). Growth of *Legionella pneumophila* in thioglycolate-elicited peritoneal macrophages from A/J mice. *Infect. Immun.* **56**(2), 370–5.

Yee, R. B. and R. M. Wadowsky (1982). Multiplication of *Legionella pneumophila* in unsterilized tap water. *Appl. Environ. Microbiol.* **43**(6), 1330–4.

CHAPTER 8

Dendritic cells, macrophages and cross-presentation of bacterial antigens: a lesson from *Salmonella*

Mary Jo Wick
Göteborg University

8.1 INTRODUCTION

Immunity to a bacterial pathogen requires the generation of bacteria-specific T cells with appropriate effector function. Eliciting T cells during infection requires internalization of the bacteria and processing of bacterial proteins to generate peptides for presentation on major histocompatibility complex (MHC) class I (MHC-I) and/or MHC class II (MHC-II) molecules, depending on the pathogen. As not all host cells have the capacity to phagocytose bacteria, and not all bacterial pathogens have the capacity to actively invade non-phagocytic cells, phagocytic antigen presenting cells, macrophages and immature dendritic cells (DCs), are the key players in generating adaptive immunity to bacteria.

Both macrophages and immature dendritic cells can present antigens from the bacteria they internalize on their own MHC-I and MHC-II molecules and thus carry out so-called direct presentation of bacterial antigens (Sundquist *et al.*, 2004; Harding *et al.*, 2003). Direct presentation of bacterial antigens on MHC-II is the expected outcome following phagocytosis of bacteria and is the event necessary to elicit $CD4^+$ T cells. However, both macrophages and DCs can also present antigens from internalized bacteria on MHC-I, molecules most renowned for their presentation of peptides derived from endogenously synthesized proteins (Rock and Goldberg, 1999), and generate $CD8^+$ T cells (Sundquist *et al.*, 2004; Harding *et al.*, 2003).

Given the capacity of both macrophages and DCs to directly present bacterial antigens on MHC-I and MHC-II, these cells in principle could initiate adaptive immunity during primary infection. However, it is only DCs that have this ability (Banchereau and Steinman, 1998). This is due to their ability to migrate to secondary lymphoid organs and interact with naive T cells in addition to their antigen presentation capacity. Dendritic cells also have sufficient expression of MHC and costimulatory molecules to activate naive T cells. Thus, DCs fulfill a unique niche in anti-bacterial immunity as the key phagocytic cells that start an adaptive immune response during infection.

This chapter summarizes data relating to presentation of antigens derived from the facultative intracellular bacteria *Salmonella enterica* serovar *typhimurium* (*S. typhimurium*) by DCs. Particular emphasis will be on antigen presentation on MHC-I. In addition to discussing direct presentation of antigens from *Salmonella*-infected cells, presentation of bacterial antigens by cells that are not themselves infected will also be discussed. That is, a non-infected DC can acquire antigenic material from other infected cells that die as a consequence of *Salmonella* in a process called indirect or cross-presentation (see also Chapter 3). Cross-presentation of bacterial antigens by bystander DCs will also be summarized. Before discussing antigen presentation, a brief overview of *Salmonella* infection will be given.

8.2 *SALMONELLA* INFECTION AND SURVIVAL IN HOST CELLS

S. typhimurium is an enteric pathogen naturally acquired orally in contaminated food or water. This Gram-negative facultative intracellular bacterium penetrates the intestinal epithelium, initially seeding Peyer's patches, and being detected thereafter in deeper tissues including the mesenteric lymph nodes, spleen and liver (Sundquist and Wick, 2005; McSorley *et al.*, 2002). In addition to M cell-mediated traversal of bacteria across the intestinal epithelium, which appears to be the predominant pathway used by invasive *Salmonella* to cross the gut barrier (Niess *et al.*, 2005; Jones *et al.*, 1994), intestinal DCs seem to also be involved in the initial traversal of *Salmonella* across the intestine (Niess *et al.*, 2005; Macpherson and Uhr, 2004; Rescigno *et al.*, 2001). This is due to their ability to sample luminal *Salmonella* by extending their dendrites between epithelial cells and transport the bacteria across the epithelial layer (Niess *et al.*, 2005; Macpherson and Uhr, 2004; Rescigno *et al.*, 2001).

Once *Salmonella* penetrate the intestinal barrier, cells of the innate immune system initiate an inflammatory response and direct the adaptive immune response (Mastroeni, 2002). *Salmonella* invasion results in infiltration of neutrophils and monocytes/macrophages, as well as DCs, to infected organs (Sundquist and Wick, 2005; Kirby *et al.*, 2001, 2002;

Mastroeni, 2002). These phagocytes recognize and bind common constituents of the bacteria, and internalize and destroy them. In addition, the phagocytes release cytokines and chemokines that recruit and activate other cells to help control the infection (Sundquist and Wick, 2005; Kirby *et al.*, 2002; Mastroeni, 2002). However, *Salmonella* can survive and replicate within phagocytes, particularly macrophages and DCs (Monack *et al.*, 2004; Sheppard *et al.*, 2003; Salcedo *et al.*, 2001; Svensson *et al.*, 2000, 2001, Mariott *et al.*, 1999; Richter-Dahlfors *et al.*, 1997). This feature makes *Salmonella* a formidable challenge for the host immune response.

8.3 PROCESSING OF *SALMONELLA* FOR DIRECT PRESENTATION ON MHC-I BY INFECTED DCs

Wild type *Salmonella* remain confined in vacuolar compartments after internalization by DCs (Brumell and Grinstein, 2004; Petrovska *et al.*, 2004; Waterman and Holden, 2003). Once inside DCs, the bacteria are processed and peptides derived from *Salmonella*-encoded proteins are presented on MHC-I (Johannson and Wick, 2004; Yrlid and Wick, 2000, 2002; Niedergang *et al.*, 2000; Svensson *et al.*, 1997, 2000; Svensson and Wick, 1999). This has been demonstrated using bone marrow-derived DCs as well as DCs isolated from the spleen, mesenteric lymph nodes or liver of naive mice infected with *Salmonella ex vivo*. Moreover, both CD8α$^+$ and CD8α$^-$ splenic DC subsets internalize *Salmonella* and process the bacteria for peptide presentation on MHC-I (Yrlid and Wick, 2002).

Direct presentation of *Salmonella* antigens on MHC-I has been studied using *Salmonella* that are internalized by DCs in a process using actin-driven cytoskeletal rearrangements (Johannson and Wick, 2004; Yrlid and Wick, 2002; Yrlid *et al.*, 2001). Studies investigating the pathway used for MHC-I presentation of *Salmonella*-encoded antigens suggest that components of the cytosolic MHC-I antigen presentation pathway are used. This occurs despite the vacuolar localization of the bacteria in DC (Brumell and Grinstein, 2004; Petrovska *et al.*, 2004; Waterman and Holden, 2003). Dendritic cells require the transporter associated with antigen processing (TAP), an ER membrane protein that translocates peptides from the cytosol into the ER in an ATP-dependent fashion, to present *Salmonella* antigens on MHC-I (Yrlid *et al.*, 2001). In addition, although not formally demonstrated for *Salmonella* but assessed for another Gram-negative vacuole-confined bacteria (*E. coli*), newly synthesized MHC-I molecules and the proteasome are also required for DC presentation of bacteria-encoded antigens (Svensson and Wick, 1999). It thus appears that the cytosolic antigen presentation machinery is also used for MHC-I presentation of bacterial antigens after phagocytic uptake of *Salmonella* by DCs.

The mechanism used by *Salmonella*-infected DCs to present bacterial antigens on MHC-I appears distinct from that used by macrophages to present *Salmonella* antigens. In the case of macrophages, the TAP transporter and proteasomes are not required (Wick and Ljunggren, 1999; Song and Harding, 1996; Wick and Pfeifer, 1996). However, more thorough characterization of the pathway(s) used for MHC-I presentation of *Salmonella* antigens by infected DCs, ideally using native bacterial antigens instead of recombinantly expressed proteins, are needed. In particular, investigating whether MHC-I presentation of *Salmonella* antigens includes a pathway where the endoplasmic reticulum inter-sects with phagosomes and results in phagosomes capable of TAP-and proteasome-dependent presentation of bacterial antigens would be informa-tive (Guermonprez *et al.*, 2003; Houde *et al.*, 2003).

8.4 MODULATING OF ANTIGEN PRESENTATION BY *SALMONELLA*

As mentioned above, *S. typhimurium* has evolved strategies to survive in phagosomal environments, which otherwise kill phagocytosed bacteria. It thus follows that the intracellular survival strategies used by *Salmonella* may influence the capacity of an infected DC to process and present bacterial antigens. This, indeed, appears to be the case. For example, the *phoP/phoQ* regulatory system, which controls the expression of over 40 genes and is involved in bacterial survival in phagosomal compartments (Groisman, 2001; Ohl and Miller, 2001), can influence the ability of infected DCs to present *Salmonella*-encoded antigens. Antigens from *Salmonella* constitutively expressing *phoP*, such that *phoP*-activated genes are locked on and *phoP*-repressed genes are off, are more efficiently presented on MHC-II after bacterial internalization by DCs (Svensson *et al.*, 2000). The effect of *phoP/phoQ* was apparent when antigen presentation was quantitated after a short (2 h) but not a longer (24 h) exposure to bacteria (Niedergang *et al.*, 2000; Svensson *et al.*, 2000). The effect of *phoP/phoQ* on antigen presentation by infected DCs required live bacteria, demonstrating that bacterial gene expression was required for the effect (Svensson *et al.*, 2000). Thus, the *phoP/phoQ* regulatory locus can influence the capacity of DCs to present *Salmonella* antigens on MHC-II during a short time frame after bacterial infection. Despite the effect of *phoP/phoQ* on presentation of a *Salmonella*-encoded antigen on MHC-II by infected DCs, no effect of this locus on presentation of *Salmonella* antigens on MHC-I was found (Svensson *et al.*, 2000). The reason underlying the different effect of *phoP/phoQ* on MHC-I and MHC-II presentation of *Salmonella* antigens has not been established.

A recent report showed that additional virulence-associated genes, in this case encoded in the *Salmonella* pathogenicity island 2 (SPI2) region, reduced the presentation of a *Salmonella*-encoded antigen on MHC-II by DCs infected with opsonized bacteria (Cheminay *et al.*, 2005). This was shown in studies where presentation of an MHC-II epitope of ovalbumin was reduced when DCs co-cultured with ovalbumin were simultaneously infected with live, wildtype *Salmonella*. The reduced MHC-II presentation required viable bacteria. Furthermore, it was most apparent when DCs were infected with *Salmonella* harboring a functional SPI2 locus and when reactive nitrogen species produced by inducible nitric oxide synthase (iNOS) occurred. The mechanism of SPI2-mediated interference with MHC-II presentation by DCs infected with opsonized *Salmonella* is not fully characterized, but may have to do with SPI2-mediated modulation of intracellular transport processes (Cheminay *et al.*, 2005).

In addition to virulence genes, other factors, such as the ease of bacterial uptake and the amount of antigen present in the bacteria that are internalized, can influence the efficiency of antigen presentation. Although the above studies demonstrating a role of *phoP/phoQ* and SPI2 in modulating direct presentation of *Salmonella* antigens by infected DCs eliminated bacterial uptake and antigen abundance as factors contributing to the observed altered antigen presentation, another study showed that directing *Salmonella* to Fcγ receptors on DCs by opsonization with *Salmonella*-specific IgG enhanced the presentation of a *Salmonella*-encoded antigen on MHC-I and MHC-II (Tobar *et al.*, 2004). In contrast to previous reports (Johannson and Wick, 2004; Yrlid and Wick, 2000, 2002; Niedergang *et al.*, 2000; Svensson *et al.*, 1997, 2000), Tobar *et al.* could only detect presentation of antigens encoded in wild type *Salmonella* when the bacteria were opsonized. Based on this they concluded that virulent *Salmonella* interferes with the capacity of DCs to process the bacteria for antigen presentation, and this could be overcome by targeting the bacteria to Fcγ receptors. However, direct evidence for active inhibition of antigen presentation by *Salmonella* was lacking. It thus remains possible that increasing antigen load in the DCs by opsonization, perhaps combined with altered intracellular trafficking when internalized as immune complexes, explains the observed effect on antigen presentation. Indeed, opsonization increases the number of bacteria per DC (Hu *et al.*, 2004; Eriksson *et al.*, 2003), and targeting antigens to Fcγ receptors on DCs increases antigen presentation in several other settings (Kalergis and Ravetch, 2002; Guyre *et al.*, 2001; Machy *et al.*, 2000; Regnault *et al.*, 1999).

8.5 DENDRITIC CELLS AS BYSTANDER ANTIGEN PRESENTING CELLS

The presentation of *Salmonella* antigens by DCs discussed above focused on direct antigen presentation by infected cells. In this pathway, DCs internalize *Salmonella*, process the bacteria and display MHC molecules containing bacterial antigens on their cell surface for recognition by T cells. In other words, the *Salmonella*-infected DCs directly process the bacteria and present bacterial antigens to T cells. However, DCs can also present bacterial antigens when they themselves are not infected by the bacteria in a process called indirect presentation or, in the case of MHC-I, cross-presentation. In indirect presentation, the DCs that present *Salmonella* antigens are not infected by the bacteria per se. Instead, the DCs are non-infected bystander cells that acquired *Salmonella* antigens by internalizing neighboring cells that have undergone death due to *Salmonella* infection. *Salmonella* expressing the type III secretion system is cytotoxic to infected cells (Monack *et al.*, 2001), and dead cells cannot productively interact with T cells (Yrlid and Wick, 2000). However, the indirect antigen presentation pathway provides a safety valve where DCs "mop up" cell debris containing *Salmonella* antigens and use this material to stimulate T cells (Yrlid and Wick, 2000).

Salmonella-induced cell death has been best studied in infected macrophages and epithelial cells (Monack *et al.*, 2001). However, *Salmonella* can also kill infected DCs by a mechanism dependent on the type III secretion system (van der Velden *et al.*, 2003; Yrlid *et al.*, 2001). Whether DCs that have undergone *Salmonella*-mediated death are also a reservoir of cell debris containing *Salmonella* antigens that can be presented by neighboring, bystander DCs is presently not known. Interestingly, the capacity to act as a bystander antigen presenting cell appears to be a unique feature of DCs, as bystander macrophages ingest *Salmonella*-induced apoptotic cells but do not present peptides from *Salmonella* antigens (Yrlid and Wick, 2000). Data suggest that macrophages compete for apoptotic material and limit the antigen available for presentation by bystander DCs (Yrlid and Wick, 2000).

The precise nature of the material in the cell debris responsible for the observed bystander presentation of *Salmonella* antigens from apoptotic macrophages is not known. However, neither peptides released into the environment that bind preformed surface MHC molecules on bystander DCs nor bacteria released into the surroundings that are subsequently phagocytosed and processed by the bystander cells account for the observed presentation (Yrlid and Wick, 2000). Additional experiments are needed to characterize the antigenic material derived from the dying, *Salmonella*-infected cells.

In the case of *Salmonella* and other intracellular bacteria, the ability of DCs to scavenge bacterial antigens from dying cells and present them to the immune system would allow detection of microbes that may otherwise be elusive (Winau *et al.*, 2004). Indeed, bystander presentation of antigenic material from cells induced to undergo apoptosis due to infection with *Mycobacterium tuberculosis* has also been shown (Schaible *et al.*, 2003). In the case of this vacuole-confined bacterium, apoptotic macrophages shuttle vesicles containing mycobacterial lipids and proteins to DCs which in turn present the material to CD1b- and MHC-I-restricted T cells, respectively (Schaible *et al.*, 2003). Similarly, *Listeria*-infected macrophages are a source of bacterial antigen that can be cross-presented on MHC-I and MHC-II by bystander DCs (Janda *et al.*, 2004; Skoberne *et al.*, 2002). Unlike mycobacteria, cross-presentation of *Listeria* antigens on MHC-I by bystander DCs did not require apoptotic antigen donor cells, but unstable bacterial translation products by infected, viable donor cells was instead important (Janda *et al.*, 2004). Apoptotic macrophages infected with the fungal pathogen *Histoplasma capsulatum* are also donors of fungal antigen for presentation on MHC-I after uptake by bystander DCs (Lin *et al.*, 2005). Thus, cross-presentation of bacterial and fungal antigens by bystander DCs, in addition to numerous examples of viral proteins that are cross-presented by DCs (Fonteneau *et al.*, 2002), has been documented. This supports a role of indirect presentation of microbial antigens by non-infected DCs as a means to elicit microbe-specific T cells during infection.

Despite the numerous studies characterizing indirect presentation of microbial antigens by bystander DCs, relatively little information is available thus far on cross-presentation of antigens from microbes in vivo during infection. Recently, however, *Salmonella*-infected tumor cells were shown to be a source of tumor antigens that are cross-presented to CD8[+] T cells and contribute to a systemic anti-tumor response in vivo (Avogadri *et al.*, 2005). Moreover, neutrophils from *Listeria*-infected mice are a substrate for bacterial antigens that are cross-presented on MHC-I by DCs, particularly non-secreted *Listeria* antigens that are otherwise directly presented by neutrophils only very inefficiently (Tvinnereim *et al.*, 2004). *Listeria*-infected neutrophils as a source of antigen for cross-priming was confirmed by showing that neutrophil-enriched cells from infected mice resulted in cross-priming after adoptive transfer into naive mice, and by the reduction in CD8$^+$ T cell priming in mice depleted of neutrophils prior to infection (Tvinnereim *et al.*, 2004).

8.6 CONCLUDING REMARKS

Dendritic cells, which are the pivitol cells in the transition from innate to adaptive immunity, can either be direct or indirect presenters of *Salmonella* antigens. Dendritic cells can also directly and indirectly present antigens from other pathogens including bacterial, fungal and viral pathogens. Dendritic cells directly present bacterial antigens to T cells upon phagocytic processing of bacteria such as *Salmonella* that does not induce their death. They can also present bacterial antigens to T cells as bystander antigen presenting cells that engulf antigenic material from neighboring cells that have undergone bacteria-induced death. Thus, despite that *Salmonella* has evolved mechanisms to (1) survive in phagocytes during infection, (2) modulate presentation of its antigens by infected DCs and (3) induce death in infected cells, the exquisite capacity of DCs to directly and indirectly present *Salmonella* antigens counters these pathogenic mechanisms to promote adaptive immunity to this formidable intracellular pathogen.

REFERENCES

Avogadri, F., Martinoli, C., Petrovska, L. *et al.* (2005). Cancer immunotherapy based on killing of *Salmonella*-infected tumor cells. *Cancer Res.* **65**, 3920–7.

Banchereau, J. and Steinman, R. M. (1998). Dendritic cells and the control of immunity. *Nature* **392**, 245–52.

Brumell, J. H. and Grinstein, S. (2004). *Salmonella* redirects phagosomal maturation. *Curr. Opin. Microbiol.* **7**, 78–84.

Cheminay, C., Möhlenbrink, A. and Hensel, M. (2005). Intracellular *Salmonella* inhibit antigen presentation by dendritic cells. *J. Immunol.* **174**, 2892–9.

Eriksson, S., Chambers, B. J. and Rhen, M. (2003). Nitric oxide produced by murine dendritic cells is cytotoxic for intracellular *Salmonella enterica* sv. *Typhimurium. Scand. J. Immunol.* **58**, 493–502.

Fonteneau, J.-F., Larsson, M. and Bhardwaj, N. (2002). Interactions between dead cells and dendritic cells in the induction of antiviral CTL responses. *Curr. Opin. Immunol.* **14**, 471–7.

Groisman, E. A. (2001). The pleiotropic two-component regulatory system PhoP PhoQ. *J. Bacteriol.* **183**, 1835–42.

Guermonprez, P., Saveanu, L., Kleijmeer, M., Davoust, J., van Endert, P., and Amigorena, S. (2003). ER-phagosome fusion defines an MHC class I cross-presentation compartment in dendritic cells. *Nature* **425**, 397–402.

Guyre, C. A., Barreda, M. E., Swink, S. L. and Fanger, M. W. (2001). Colocalization of FcγRI-targeted antigen with class I MHC: implications for antigen processing. *J. Immunol.* **166**, 2469–78.

Harding, C. V., Ramachandra, L. and Wick, M. J. (2003). Interaction of bacteria with antigen presenting cells: influences on antigen presentation and antibacterial immunity. *Curr. Opin. Immunol.* **15**, 112–19.

Houde, M., Bertholet, S., Gagnon, E., Brunet, S., Goyette, G., Laplante, A. *et al.* (2003). Phagosomes are competent organelles for antigen cross-presentation. *Nature* **425**, 402–6.

Hu, P. Q., Tuma-Warrino, R. J., Bryan, M. A., Mitchell, K. G., Higgins, D. E., Watkins, S. C. and Salter, R. D. (2004). *Escherichia coli* expressing recombinant antigen and listeriolysin O stimulate class I-restricted CD8+ T cells following uptake by human APC. *J. Immunol.* **172**, 1595–601.

Janda, J., Schöneberger, P., Skoberne, M., Messerle, M., Rüssmann, H., and Geginat, G. (2004). Cross-presentation of *Listeria*-derived CD8 T cell epitopes requires unstable bacterial translation products. *J. Immunol.* **173**, 5644–51.

Johannson, C. and Wick, M. J. (2004). Liver dendritic cells present bacterial antigens and produce cytokines upon *Salmonella* encounter. *J. Immunol.* **172**, 2496–503.

Jones, B. D., Ghori, N. and Falkow, S. (1994). *Salmonella typhimurium* initiates murine infection by penetrating and destroying the specialized epithelial M cells of the Peyer's patches. *J. Exp. Med.* **180**, 15–23.

Kalergis, A. M. and Ravetch, J. V. (2002). Inducing tumor immunity through the selective engagement of activating Fcγ receptors on dendritic cells. *J. Exp. Med.* **195**, 1653–9.

Kirby, A. C., Yrlid, U. and Wick, M. J. (2002). The innate immune response differs in primary and secondary *Salmonella* infection. *J. Immunol.* **169**, 4450–9.

Kirby, A. C., Yrlid, U., Svensson, M. and Wick, M. J. (2001). Differential involvement of dendritic cell subsets during acute *Salmonella* infection. *J. Immunol.* **166**, 6802–11.

Lin, J.-S., Yang, C.-W., Wang, D.-W. and Wu-Hsieh, B. A. (2005). Dendritic cells cross-present exogenous fungal antigens to stimulate a protective CD8 T cell response in infection by *Histoplasma capsulatum*. *J. Immunol.* **174**, 6282–91.

Machy, P., Serre, K. and Leserman, L. (2000). Class I-restricted presentation of exogenous antigen acquired by Fcγ receptor-mediated endocytosis is regulated in dendritic cells. *Eur. J. Immunol.* **30**, 848–57.

Macpherson, A. J. and Uhr, T. (2004). Induction of protective IgA by intestinal dendritic cells carrying commensal bacteria. *Science* **303**, 1662–5.

Mariott, I., Hammond, T. G., Thomas, E. K. and Rost, K. L. (1999). *Salmonella* efficiently enter and survive within cultured CD11c$^+$ dendritic cells initiating cytokine expression. *Eur. J. Immunol.* **29**, 1107–15.

Mastroeni, P. (2002). Immunity to systemic *Salmonella* infections. *Curr. Molec. Med.* **2**, 393–406.

McSorley, S. J., Asch, S., Costalonga, M., Reinhardt, R. L. and Jenkins, C. (2002). Tracking *Salmonella*-specific CD4 T cells in vivo reveals a local mucosal response to a disseminated infection. *Immunity* **16**, 365–77.

Monack, D. M., Navarre, W. W. and Falkow, S. (2001). *Salmonella*-induced macrophage death: the role of caspase-1 in death and inflammation. *Microb. Infect.* **3**, 1201–12.

Monack, D. M., Bouley, D. M. and Falkow, S. (2004). *Salmonella typhimurium* persists within macrophages in the mesenteric lymph nodes of chronically infected *Nramp1*$^{+/+}$ mice and can be reactivated by IFNγ neutralization. *J. Exp. Med.* **199**, 231–41.

Niedergang, F., Sirard, J.-C., Tallichet Blanc, C. and Kraehenbuhl, J.-P. (2000). Entry and survival of *Salmonella typhimurium* in dendritic cells and presentation of recombinant antigens do not require macrophage-specific virulence factors. *Proc. Natl Acad. Sci. U S A* **97**, 14650–5.

Niess, J. H., Brand, S., Gu, X., Landsman, L., Jung, S., McCormick, B. A. *et al.* (2005). CX$_3$CR1-mediated dendritic cell access to the intestinal lumen and bacterial clearance. *Science* **307**, 254–8.

Ohl, M. E. and Miller, S. I. (2001). *Salmonella*: A model for bacterial pathogenesis. *Annu. Rev. Med.* **52**, 259–74.

Petrovska, L., Aspinall, R. J., Barber, L., Clare, S., Simmons, C. P., Stratford, R. *et al.* (2004). *Salmonella enterica* serovar *Typhimurium* interaction with dendritic cells: impact of the *sifA* gene. *Cellular Microbiol.* **6**, 1071–84.

Regnault, A., Lankar, D., Lacabanne, V., Rodriguez, A., Théry, C., Rescigno, M. *et al.* (1999). Fcγ receptor-mediated induction of dendritic cell maturation and major histocompatibility complex class I-restricted antigen presentation after immune complex internalization. *J. Exp. Med.* **189**, 371–80.

Rescigno, M., Urbano, M., Valzasina, B., Francolini, M., Rotta, G., Bonasio, R. *et al.* (2001). Dendritic cells express tight junction proteins and penetrate gut epithelia monolayers to sample bacteria. *Nature Immunol.* **2**, 361–7.

Richter-Dahlfors, A., Buchan, A. M. J. and Finlay, B. B. (1997). Murine salmonellosis studied by confocal microscopy: *Salmonella typhimurium* resides intracellularly inside macrophages and exerts a cytotoxic effect on phagocytes in vivo. *J. Exp. Med.* **186**, 569–80.

Rock, K. L. and Goldberg, A. L. (1999). Degradation of cell proteins and the generation of MHC class I-presented peptides. *Annu. Rev. Immunol.* **17**, 739–79.

Salcedo, S. P., Noursadeghi, M., Cohen, J. and Holden, D. W. (2001). Intracellular replication of *Salmonella typhimurium* strains in specific subsets of splenic macrophages in vivo. *Cell. Microbiol.* **3**, 587–97.

Schaible, U. E., Winau, F., Sieling, P. A., Fischer, K., Collins, H. L., Hagens, K. *et al.* (2003). Apoptosis facilitates antigen presentation to T lymphocytes through MHC-I and CD1 in tuberculosis. *Nature Med.* **9**, 1039–46.

Sheppard, M., Webb, C., Heath, F., Mallos, V., Emilianus, R., Maskell, D., and Mastroeni, P. (2003). Dynamics of bacterial growth and distribution within the liver during *Salmonella* infection. *Cellular Microbiol.* **5**, 593–600.

Skoberne, M., Schenk, S., Hof, H. and Geginat, G. (2002). Cross-presentation of *Listeria monocytogenes*-derived CD4 T cell epitopes. *J. Immunol.* **169**, 1410–18.

Song, R. and Harding, C. V. (1996). Roles of proteasomes, transporter for antigen presentation (TAP), and β2-microglobulin in the processing of bacterial or particulate antigens via an alternate class I MHC processing pathway. *J. Immunol.* **156**, 4182–90.

Sundquist, M. and Wick, M. J. (2005). TNF-α-dependent and-independent maturation of dendritic cells and recruited CD11c^{int}CD11b$^+$ cells during oral *Salmonella* infection. *J. Immunol.* **175**, 3287–98.

Sundquist, M., Rydström, A. and Wick, M. J. (2004). Immunity to *Salmonella* from a dendritic point of view. *Cell. Microbiol.* **6**, 1–11.

Svensson, M. and Wick, M. J. (1999). Classical MHC-I presentation of a bacterial fusion protein by bone marrow-derived dendritic cells. *Eur. J. Immunol.* **29**, 180–8.

Svensson, M., Stockinger, B. and Wick, M. J. (1997). Bone marrow-derived dendritic cells can process bacteria for MHC-I and MHC-II presentation to T cells. *J. Immunol.* **158**, 4229–36.

Svensson, M., Johansson, C. and Wick, M. J. (2000). *Salmonella enterica* serovar *Typhimurium*-induced maturation of bone marrow-derived dendritic cells. *Infect. Immun.* **68**, 6311–20.

Svensson, M., Johannson, C. and Wick, M. J. (2001). *Salmonella typhimurium*-induced cytokine production and surface molecule expression by murine macrophages. *Microb. Pathog.* **31**, 91–102.

Tobar, J. A., González, P. A. and Kalergis, A. M. (2004). *Salmonella* escape from antigen presentation can be overcome by targeting bacteria to Fcγ receptors on dendritic cells. *J. Immunol.* **173**, 4058–65.

Tvinnereim, A. R., Hamilton, S. E. and Harty, J. T. (2004). Neutrophil involvement in cross-priming CD8[+] T cell responses to bacterial antigens. *J. Immunol.* **173**, 1994–2002.

van der Velden, A. W. M., Velassquez, M. and Starnbach, M. N. (2003). *Salmonella* rapidly kill dendritic cells via a caspase-1-dependent mechanism. *J. Immunol.* **171**, 6742–9.

Waterman, S. R. and Holden, D. W. (2003). Functions and effectors of the *Salmonella* pathogenicity island 2 type III secretion system. *Cellular Microbiol.* **5**, 501–11.

Wick, M. J. and Pfeifer, J. D. (1996). MHC-I presentation of OVA(257–264) from exogenous sources: protein context influences the degree of TAP-independent presentation. *Eur. J. Immunol.* **26**, 2790–9.

Wick, M. J. and Ljunggren, H.-G. (1999). Processing of bacterial antigens for peptide presentation on MHC class I molecules. *Immunol. Rev.* **172**, 153–62.

Winau, F., Kaufmann, S. H. E. and Schaible, U. E. (2004). Apoptosis paves the detour path for CD8 T cell activation against intracellular bacteria. *Cellular Microbiol.* **6**, 599–607.

Yrlid, U. and Wick, M. J. (2000). *Salmonella*-induced apoptosis of infected macrophages results in presentation of a bacteria-encoded antigen after uptake by bystander dendritic cells. *J. Exp. Med.* **191**, 613–23.

Yrlid, U. and Wick, M. J. (2002). Antigen presentation capacity and cytokine production by murine splenic dendritic cell subsets upon *Salmonella* encounter. *J. Immunol.* **169**, 108–16.

Yrlid, U., Svensson, M., Kirby, A. C. and Wick, M. J. (2001). Antigen-presenting cells and anti-*Salmonella* immunity. *Microb. Infect.* **3**, 1239–48.

PART IV   Dendritic cells and immune
evasion of bacteria in vivo

CHAPTER 9

Pathogen-recognition receptors as targets for pathogens to modulate immune function of antigen-presenting cells

Anneke Engering, Sandra J. van Vliet, Estella A. Koppel, Teunis B. H. Geijtenbeek and Yvette van Kooyk
VU Medical Center

9.1 INTRODUCTION

Antigen-presenting cells (APC), such as dendritic cells (DCs) and macrophages, are located throughout the body to sense and capture invading pathogens and to trigger immune responses to fight such invaders. In addition, in the absence of danger signals, DCs have an active role in the induction of T cell tolerance and the maintenance of homeostasis. The recognition and internalization of pathogens is mediated by so-called pathogen-recognition receptors, germ-line encoded cell surface receptors that include toll-like receptors (TLR) and C-type lectins (CLR). It is becoming increasingly clear that during the long co-evolution with their hosts, pathogens have evolved mechanisms to misuse pathogen-recognition receptors to suppress or evade immune responses and thus to escape clearance. In this chapter, we will review recent examples of how pathogens evade immune activation by targeting recognition receptors on APC and subverting their function.

9.2 BACTERIAL RECEPTORS ON ANTIGEN-PRESENTING CELLS

APC interact with invading pathogens via pathogen-recognition receptors that bind conserved patterns of carbohydrates, lipids, proteins and nucleic acids in classes of microbes[1,2]. This variety of receptors and conserved ligands recognized ensures that most, if not all, microbes can be detected by the immune system, either by a single or by combinations of receptors. Pathogen-recognition receptors include TLR[3] and CLR[4] (Figure 9.1). To date, 11 TLR have been identified (see Chapter 2) that each targets specific pathogenic structures, such as lipopolysaccharide (TLR4), viral

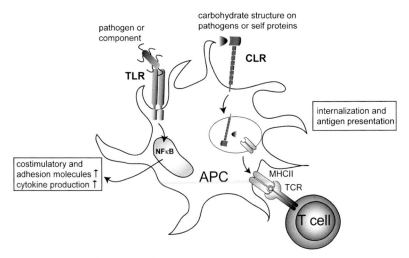

Figure 9.1. TLR and CLR are pathogen-recognition receptors on APC. APC express several classes of receptors that mediate recognition and internalization of pathogens, including TLR and CLR. Recognition of pathogens or their components by TLR leads to activation of intracellular signalling pathways, including NFκB translocation, and subsequent upregulation of expression of costimulatory and adhesion molecules and production of cytokines. CLR bind carbohydrate structures on pathogens resulting in internalization, and intracellular processing for presentation by MHC II molecules to T cells. TCR, T cell receptor.

dsRNA (TLR3) and bacterial peptidoglycans (TLR2/TLR6)[5]. Upon inter-action with a pathogen, TLR transmit this information through signalling pathways resulting in activation of APC, including expression of costimu-latory molecules and production of inflammatory cytokines (Figure 9.1). APC react differently to distinct microbial ligands, even though there is an overlap of adaptor molecules and signalling routes between different TLR. Indeed, recent studies support the hypothesis that different TLR ligands activate distinct downstream responses leading to tailored activation of APC to most effectively fight the specific pathogen[6].

Whereas TLR are instrumental for alerting the immune system for invading pathogens, other pathogen receptors on APC play a role in capturing the microbes for intracellular degradation and antigen presenta-tion. These include the family of CLR that recognize pathogens by their carbohydrate structures[7]. However, the range of ligands that interact with CLR include not only pathogenic structures but also self molecules, mediating cell−cell adhesion as well as internalizing endogenous ligands for homeostatic control[8]. Binding to carbohydrates is mediated by the carbohydrate-recognition domain (CRD) of the CLR[9]. The more than 15 CLR that have been cloned from DC and macrophages so far either contain a single CRD, including DC-SIGN, MGL, and Dectin-1, or multiple CRDs, such as MR (8 CRDs) and DEC-205 (10 CRDs). The exact carbohydrate moieties recognized by the distinct CLR are currently being explored and depend not only on the type of glycan (such as mannose, fucose or galactose) but also on the complexity, multivalency, and branching of the carbohydrates (see http://web.mit.edu/glycomics/consortium/).

Most CLR are highly expressed on immature DCs and function as endocytic receptors that capture antigens for presentation (Figure 9.1). Distinct internalization motifs present in the cytoplasmic part of several CLR can mediate and guide endocytosis and intracellular trafficking[4]. For example, ligand binding to DC-SIGN triggers internalization into lysosomal compartments, whereas MR continuously recycles between the cell surface and early endosomes, where ligand is released[10–12]. Overall, such receptor-mediated antigen uptake results in high amounts of internalized antigens for antigen processing and presentation to T cells. However, several pathogens including HIV-1 have evolved to subvert these internalization routes resulting in poor antigen-presentation (see below). In addition to endocytic motifs, some CLR contain putative signalling motifs including immunoreceptor tyrosine-based inhibitory (ITIM) and activatory (ITAM) motifs, indicating that such CLR have immuno-suppressive or activatory functions[4].

Other families of pathogen-recognition receptors include the cyto-plasmic surveillance proteins NOD1 and NOD2, scavenger receptors, other lectin receptors (e.g. galectins) and opsonic receptors for immunoglobulins (FcR) and complement[1,13]. Each of these receptors can potentially trigger intracellular signals and thus contribute to the final outcome of the immune response. On the other hand, during co-evolution of pathogens with their hosts, each of these receptors could have provided means for pathogens to modulate, or even evade, immune responses.

9.3 PATHOGENS ARE RECOGNIZED BY COMBINATIONS OF RECEPTORS

Although many studies have focused on isolated ligands for pathogen-recognition receptors, in the body APC will mainly encounter intact pathogens that contain a variety of potential recognition elements. In addition, during destruction of pathogens upon internalization into the endosomal/lysosomal pathway, alternative ligands can become available for triggering of intracellularly expressed receptors. TLR9, for instance, is localized intracellularly in endosomes and recognizes DNA derived from internalized and degraded bacteria[14]. Thus, whole microbes will be recognized by combinations of distinct receptors both on the cell surface and intracellular. Simultaneous and/or sequential activation of multiple innate receptors will assist in tailoring an effective response to a specific pathogen. Besides intact pathogens that are recognized during initial infection, at later stages and during chronic infection soluble microbial components can be secreted that bind to recognition receptors on APC. For example *Mycobacterium tuberculosis* infects primarily macrophages, but at later stages after infection induces secretion of its cell wall component mannosylated lipoarabinomannan (ManLAM), that modulates DC function (see below).

Moreover, different APC express distinct combinations of pathogen-recognition receptors; both the amounts and the relative abundance can vary. For example, whereas DCs express the CLR DC-SIGN and MR and low amounts of TLR1 and TLR2, monocytes/macrophages express MR, but not DC-SIGN, and high levels of TLR1 and TLR2[15,16]. This will result in distinct immune effector responses depending on the subset of APC that encounters a pathogen. Also, the availability and the affinity of the receptors will determine whether they are triggered or not. Selective expression of pathogen-recognition receptors in defined intracellular compartments together with the route of the pathogen upon internalization will govern ligand interactions. In addition, receptors can have overlapping ligand specificity but may differ in affinity for pathogenic component, thus resulting in dominance of one receptor for interacting with a specific pathogen. Detailed knowledge on expression, ligand specificity and affinity of recognition receptors as well as on their signalling capacities is required to decipher the immune response triggered by an intact pathogen.

9.4 BACTERIA CAN USE TLR TO EVADE IMMUNE RESPONSES

TLR play a crucial role in alarming the immune system upon invasion of pathogens[3]. Upon ligation of TLR, DCs are activated and migrate from the sites of infection to the lymph nodes, for antigen presentation to T cells. Thus, DCs bridge innate and acquired immune responses. The importance of TLR became clear in mice and humans that lack specific TLR or intracellular signalling components, leading to an increased susceptibility to a variety of microbes. In addition, such models have identified specific ligands for each TLR; for example products from Gram-positive bacteria are recognized by TLR2, whereas TLR4 mediates recognition of Gram negative bacteria and their components. However, besides activation of efficient immune responses, recent findings show that certain pathogens misuse TLR to escape the host defense[17].

Several bacteria have found a way to control inflammation by modification of TLR ligands. In general, TLR recognize conserved microbial compounds that are essential for pathogenicity and survival of pathogens. However, it is becoming clear that modifications of TLR ligands do occur between species and even within one species, thereby altering the interaction and subsequent signalling of TLR. One example is lipid A, the part of lipopolysaccharide that is recognized by TLR4 and that is an essential component of the outer membrane of Gram-negative bacteria. Whereas *E. coli*-derived lipid A is a strong activator of TLR4, lipid A from other bacteria can be modified resulting in low affinity (e.g. LPS from *Salmonella*[18]) or even escape from recognition by TLR4 (e.g. LPS from *P. gingivalis*[19]). In addition to modification of TLR ligands, pathogens can block specific components of the TLR signalling pathway, thereby reducing activation of APC and immune responses.

Other bacteria have found means to directly induce immune suppression instead of immune activation by signalling through TLR2. In comparison to TLR4 activation, TLR2 ligation induces weak inflammatory responses and strong anti-inflammatory effects[20]. This is probably part of the recovery phase that occurs after inflammation and clearance to allow the body to return to normal steady-state conditions. Recent studies have shown that pathogens exploit this anti-inflammatory effect of TLR2 to avoid inflammation and thus promote survival[20] (Figure 9.2b). Examples include pathogenic components that solely trigger TLR2, such as phosphatidyl serine derived from *Schistosoma* that activate DCs to induce regulatory T cells[21], or bacteria that reduce TLR4 triggering in favor of TLR2 activation like *P. gingivalis*[19] and *Yersinia* that induces immunosuppression via IL-10[22]. However, it is not clear if TLR2 is the only receptor that is involved in the described immunosuppression or that members of other pathogen receptor families are simultaneously triggered, such as CLR (see below).

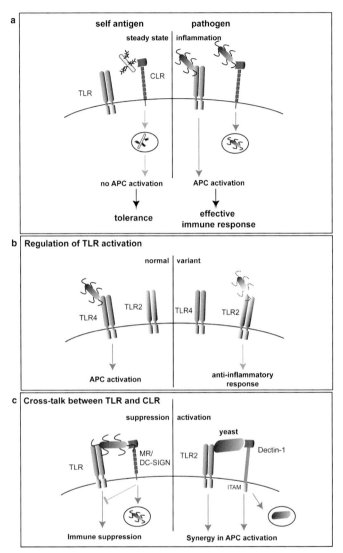

Figure 9.2. Pathogens are recognized by combinations of receptors, but can also target these receptors to escape immune responses. (a) Self-antigens that bind to CLR in steady-state conditions can be internalized by APC, but no activated occurs resulting in induction of tolerance. However, upon simultaneous recognition of pathogens by CLR and TLR, APC become activated and effective immune responses are induced. (b) Certain pathogens, such as *P. gingivalis*, can modify its TLR ligands to induce immune suppression by reducing TLR4 triggering in favor of TLR2 activation. (c) Cross-talk between TLR and CLR can either result in immune suppression or in synergistic activation of APC. Certain ligands of the MR and DC-SIGN induce down modulation of TLR signalling and inhibition of immune activation, whereas collaborative signalling of Dectin-1 and TLR2 upon binding of yeast induces enhanced activation of APC.

9.5 TARGETING TO CLR CAN INDUCE EITHER TOLERANCE OR IMMUNE ACTIVATION

In contrast to TLR, many CLR were initially identified as receptors for endogenous ("self") ligands. The MR plays a role in the removal of self glycoproteins from the circulation, whereas MR, DC-SIGN and Dectin-1 are involved in cell−cell interactions[23−27]. In addition, triggering of CLR in the absence of inflammatory stimuli does not lead to immune activation but rather to unresponsiveness. This was first shown by targeting antigens to DEC-205 on immature DCs, resulting in tolerance and the deletion of effector T cells[28]. This leads to the hypothesis that the physiological function of these receptors is the recognition of glycosylated self-antigens for homeostatic control and that pathogens have misused CLR to escape clearance (Figure 9.2a). However, recent evidence indicates that some CLR, such as Dectin-1, function as a pathogen receptor that activates APC. Several studies of pathogens that simultaneously trigger CLR and TLR will be discussed below, showing that cross-talk between CLR and TLR can either lead to suppression or to activation of immune responses (Figure 9.2c).

9.6 CROSS-TALK BETWEEN CLR AND TLR CAN RESULT IN IMMUNE SUPPRESSION

A recent study shows that targeting MR with an activating anti-MR antibody or with certain natural ligands could prime a regulatory program in DCs, leading to the production of anti-inflammatory cytokines and induction of Th2 cells with regulatory capacity[29]. MR triggering could also prevent the production of inflammatory cytokines by the TLR4-ligand LPS implying cross-talk between the MR and TLR4 leading to down modulation of TLR4 signalling (Figure 9.2c). Interestingly, only a restricted set of ligands, such as mycobacterial-derived ManLAM and a complex proteoglycan could prime regulatory DCs, whereas other ligands including mannan and dextran, had no effect on DCs[29]. This differential effect could be due to the fact that several of these ligands can also bind to other C-type lectins, including DC-SIGN, and might thus influence activation of DCs.

DC-SIGN is a DC-specific CLR that binds self-glycoproteins as well as a broad range of pathogens through high mannose and Lewis-containing carbohydrates (Figure 9.3). Although DC-SIGN has been shown to mediate internalization of ligands for antigen presentation[10], certain pathogens, such as HIV and HCV, induce misrouting after internalization through

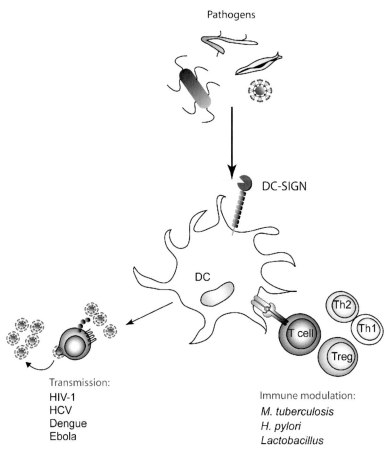

Figure 9.3. Pathogens target DC-SIGN to escape immune responses. Immature DC express CLR such as DC-SIGN that recognizes different pathogens through their carbohydrate structures. The immunological outcome of this interaction is specific for the pathogen and also depends on co-ligation of other pathogen receptors, such as TLR. Viruses can use DC-SIGN for transmission to T cells, whereas several bacteria can modulate DC function through DC-SIGN.

DC-SIGN to non-lysosomal compartments, allowing escape from degradation and antigen presentation[30,31]. In this way, a part of the internalized HIV remains virulent and can infect T cells *in trans* upon migration of DCs to lymph nodes[32,33]. In addition to HIV, DC-SIGN binds and internalizes Dengue virus, HCMV, HCV and Ebola for *in-trans* infection of target cells[34–36] (Figure 9.3).

In addition to viruses, other pathogens also target DC-SIGN, including mycobacteria, yeast and schistosoma parasites[37–39]. Recognition of such whole pathogens relies not only on DC-SIGN, but also on other pathogen receptors. We will highlight three examples of bacteria that simultaneously trigger DC-SIGN and other pathogen-recognition receptors to modulate DC function (Figure 9.2c and Figure 9.3).

Tuberculosis, caused by *Mycobacterium tuberculosis*, is one of the major infectious diseases worldwide. Although *M. tuberculosis* primarily infects macrophages, it also binds DC-SIGN on DCs through interactions with ManLAM, a major component of its cell wall that can be secreted by infected cells[40,41]. Recent studies have demonstrated that ManLAM binding to DC-SIGN blocks TLR4-induced maturation of DCs and induces production of the anti-inflammatory cytokine IL-10[40]. This indicates that binding of ManLAM to DC-SIGN triggers inhibitory signals that may enable pathogens such as *M. tuberculosis* to suppress immune activation signals through TLR.

The human gastric pathogen *Helicobacter pylori* causes persistent infection in half of mankind, which requires a certain balance between mild inflammation and escape of clearance. *H. pylori* has evolved many complex mechanisms to subvert innate and adaptive immunity in its host, such as resistance to phagocytic killing, inhibition of antigen processing and suppression of T cell activation[42]. Recently, we have shown that differential targeting of DC-SIGN on DCs provides *H. pylori* with yet another mechanism to modulate immune responses[43]. *H. pylori* LPS express Lewis antigens that are subject to phase variation, meaning that the genes responsible for these epitopes are switched on and off with high frequency[44]. Lewis phase variation is caused by frame shifts in glycosyltransferase genes and results in LPS with and without Lewis carbohydrates on bacteria within a single strain. Although both variants trigger TLR2 and TLR4 to a similar extent, these modifications result in a differential activation to DCs through targeting of DC-SIGN[43]. *H. pylori* that express Lewis-containing LPS bind to DC-SIGN and inhibit Th1-polarization compared to Lewis-negative variants that are not recognized by DC-SIGN. The Lewis-negative and -positive strains are identical except for the presence of Lewis carbohydrates on LPS, implying that *H. pylori* uses phase variation to regulate targeting to DC-SIGN and to modulate DC function. In addition to *H. pylori* other pathogens that target DC-SIGN are also likely to block Th1 skewing, such as *Schistosoma mansoni* and *Leishmania* species that both favor Th2, and not Th1, responses for chronic infections.

Besides these Th1/Th2 modulatory effects of DC-SIGN, we have found that DC-SIGN targeting by *Lactobacillus* can prime mature regulatory DCs that induce regulatory T cell differentiation[45]. Lactobacilli are one of the most frequently used probiotics, i.e. live microbial food ingredients that are beneficial to health. Whereas several lactobacilli bound to DCs, only *L. reuteri* and *L. casei* that target DC-SIGN could prime the development of regulatory T cells. Anti-DC-SIGN antibodies abrogated this suppressor activity, indicating a crucial role for DC-SIGN herein. Interestingly, addition of maturation stimuli was needed for full regulatory T cell-inducing capacity of DCs by lactobacilli, suggesting that cross-regulation between DC-SIGN and other signalling pathways occurs.

Interestingly, mSIGNR1, the murine homologue of DC-SIGN, is not expressed on DCs but on subsets of macrophages in lymph nodes, spleen and liver at locations important in the defence against pathogens[46]. Several pathogens have been identified to interact with mSIGNR1 in vitro, such as the virus HIV-1[47], the yeast *C. albicans*[48], the mycobacteria *M. tuberculosis* and *M. bovis* BCG[40,48] as well as bacteria including *Streptococcus pneumoniae*, *E. coli* and *S. typhimurium*[49,50]. The importance of mSIGNR1 during infection was demonstrated using mice deficient for mSIGNR1; such mice are more susceptible to *S. pneumoniae* challenge[51]. In contrast to DC-SIGN, that suppresses or modulates TLR signalling, mSIGNR1 was recently demonstrated to enhance TLR4-mediated cytokine production by macrophages upon binding to core polysaccharides of LPS[52]. mSIGNR1 lacks signalling motifs in its cytoplasmic tail, but its association with TLR4 might be involved in the observed synergistic effect[52]. The generation of knockout and double-knockout mice will enable further insight in the function of mSIGNR1 as well as cross-talk with TLR.

9.7 CROSS-TALK BETWEEN CLR AND TLR CAN RESULT IN IMMUNE ACTIVATION

Another example of cross-talk between CLR and TLR signalling was shown recently for Dectin-1 and TLR2 that synergize to activate inflammatory responses. Dectin-1 is a CLR with a single CRD that binds β-glucans, such as present in yeast, and is expressed by DCs and macrophages[53]. In its cytoplasmic tail, Dectin-1 contains an activatory tyrosine-based motif (ITAM); ligand binding induces tyrosine phosphorylation. In addition to Dectin-1, yeast particles also trigger signalling through TLR2/6 hetero-dimers resulting in an inflammatory response[54]. Two recent papers have shown that upon simultaneous recognition of heat-killed yeast by Dectin-1 and TLR2/6 collaborative signalling of these two pathogen receptors induces enhanced inflammatory cytokine production by APC[55,56]. Dectin-1 is not only involved in binding of yeast through β-glucans, it also mediates phagocytosis, whereas TLR are not involved in this process[54]. Strikingly, the signalling pathway of Dectin-1 differs from the TLR pathway, but also from other phagocytic receptors such as FcR[57], and involves phosphorylation of the cytoplasmic tyrosine and recruitment of Syk kinase[58].

Interestingly, besides the CLR Dectin-1, several other CLR can also bind yeast, including MR and DC-SIGN[37,39], although the glycans that are recognized differ. DC-SIGN and the MR bind yeast mannan, whereas Dectin-1 interacts with β-glucans. Moreover, unicellular yeast is differentially recognized from the virulent filamentous form (hyphae) of *Candida* by DC. β-Glucans, which are recognized by Dectin-1, are normally shielded on live yeasts by the outer wall components. The normal mechanism of yeast budding and cell separation creates scars that expose enough β-glucans to trigger Dectin-1 mediated phagocytosis. During filamentous growth no cell separation and therefore no scarring occurs and the pathogen fails to activate Dectin-1[59]. Other studies have shown that DCs induce protective Th1 responses upon unicellular yeast phagocytosis but pathology and Th2 induction after hyphae uptake, although the contribution of the different receptors in such differentiation remains unclear[60,61]. To fully understand the immune response triggered by whole yeast, the complex interplay between multiple receptors involved in recognition has to be further unraveled. Comparison of gene expression profiles of DCs in response to *Candida albicans* or isolated yeast mannan revealed that although a common gene program is induced by both compounds, mannan activated an additional set of genes[62]. This implies that recognition is indeed different between whole yeast and its cell-wall component mannan and results in differential activation of DCs.

9.8 OTHER EXAMPLES OF CROSS-TALK BETWEEN PATHOGEN RECEPTORS

The complexity of recognition of pathogens and their components was emphasized in a recent study on outer membrane protein A (OmpA) of *Klebsiella*[63]. Whereas recognition by macrophages and DCs is mediated by the scavenger receptors Lox-1 and SREC-1, TLR2 triggering results in activation, including production of the soluble pathogen-recognition receptor PTX3. Subsequently, PTX3 binds OmpA and enhances inflammatory responses[63]. Immune responses to whole *Klebsiella* will probably be the result of triggering not only Lox-1, SREC-1, TLR-2 and PTX3 that recognize OmpA, but additional receptors recognizing other *Klebsiella* components that may function complementary or contradictory.

In macrophages, synergy was reported between TLR 2, 4, 7 and 9 and the receptor for adenosine, a product of tissue damage/inflammation. Concomitant stimulation results in increased production of vascular endothelial growth factor and downregulation of TNF-α[64]. Such synergy could play a role in tissue repair and resolving inflammation after clearance of the pathogen.

Depending on the stimuli received macrophages can adopt different modes of action[65,66]. Classically activated macrophages develop after IFNγ activation along with exposure to a microbe or a microbial product such as LPS. These cells are easily identified via their production of high levels of nitric oxide (NO) and reactive oxygen species (ROS). Classically activated macrophages produce high amounts of IL-12 and stimulate Th1 responses. They also possess an enhanced ability to kill intracellular pathogens. If however TLR activation is accompanied by FcR triggering via immune complexes, macrophages differentiate into type II, alternatively activated macrophages. These macrophages no longer produce IL-12; instead they secrete high amounts of IL-10. By virtue of this IL-10 production, type II-activated macrophages have a strong anti-inflammatory function and preferentially induce Th2-type responses. Upon parasitic infections, expression of the galactose-type CLR MGL2 is induced on murine alternatively activated macrophages[67]. Interestingly, mMGL2 is highly homologous to human MGL[68] that was recently shown to bind glycan structures containing terminal GalNAc moieties, expressed among others by the human helminth parasite *Schistosoma mansoni*[69], pointing to a role for MGL as a pathogen receptor.

9.9 CONCLUDING REMARKS

As discussed above, recent advances indicate that the balance between pathogen-recognition receptors upon encountering of a pathogen by APC will determine the outcome of an immune response, either immune activation or suppression. In order to fully understand the impact of pathogen recognition, not only the ligand specificity and subsequent signalling pathway needs to be elucidated, but also studies on highly defined pathogenic components and/or molecularly modified pathogens, on different DCs and macrophage subsets and in animal models are required.

On the other hand, information on the pathogen itself and variations are crucial to be able to comprehend immune responses.

For several CLR, the outcome of the subsequent immune response against a pathogen depends on the context in which the ligand is recognized (i.e. activation of APC through other receptors). In case of self-glycoproteins that bind to CLR but do not activate APC, immune tolerance will be generated, whereas pathogens that bind to CLR and strongly activate TLR will induce antigen presentation to T cells by fully activated DCs and generation of protective immune responses. However, pathogens that down modulate TLR activation or provide a dominant signal through specific CLR, such as DC-SIGN and/or MR, can inhibit immune activation to promote survival. Such information could be used to suppress unwanted inflammatory responses as in autoimmunity, for example by targeting antigens to specific CLR on DCs without additional triggers for tolerance induction or to TLR2 to induce anti-inflammatory processes. Likewise, lessons could be learned from triggering combinations of pathogen receptors that result in synergistic inflammatory effects, in order to enhance immune responses to tumor antigens or for use in vaccine development.

REFERENCES

1. Taylor, P. R., L. Martinez-Pomares, M. Stacey, H. H. Lin, G. D. Brown, and S. Gordon (2005). Macrophage receptors and immune recognition. *Annu. Rev. Immunol.* **23**, 901–44.
2. Sousa, Reis E. (2004). Activation of dendritic cells: translating innate into adaptive immunity. *Curr. Opin. Immunol.* **16**, 21–5.
3. Akira, S., K. Takeda, and T. Kaisho (2001). Toll-like receptors: critical proteins linking innate and acquired immunity. *Nat. Immunol.* **2**, 675–80.
4. Figdor, C. G., Y. van Kooyk, and G. J. Adema (2002). C-type lectin receptors on dendritic cells and Langerhans cells. *Nature Rev. Immunol.* **2**, 77–84.
5. Underhill, D. M. and A. Ozinsky (2002). Toll-like receptors: key mediators of microbe detection. *Curr. Opin. Immunol.* **14**, 103–10.
6. McGettrick, A. F. and L. A. O'Neill (2004). The expanding family of MyD88-like adaptors in Toll-like receptor signal transduction. *Mol. Immunol.* **41**, 577–82.
7. van Kooyk, Y., A. Engering, A. N. Lekkerkerker, I. S. Ludwig, and T. B. Geijtenbeek (2004). Pathogens use carbohydrates to escape immunity induced by dendritic cells. *Curr. Opin. Immunol.* **16**, 488–93.

8. Geijtenbeek, T. B., S. J. van Vliet, A. Engering, B. A. 't Hart, and Y. van Kooyk (2004). Self- and nonself-recognition by C-type lectins on dendritic cells. *Annu. Rev. Immunol.* **22**, 33–54.

9. Drickamer, K. (1999). C-type lectin-like domains. *Curr. Opin. Struct. Biol.* **9**, 585–90.

10. Engering, A., T. B. Geijtenbeek, S. J. van Vliet, M. Wijers, E. van Liempt, N. Demaurex, A. Lanzavecchia, J. Fransen, C. G. Figdor, V. Piguet, and Y. van Kooyk (2002). The dendritic cell-specific adhesion receptor DC-SIGN internalizes antigen for presentation to T cells. *J. Immunol.* **168**, 2118–26.

11. Tan, M. C., A. M. Mommaas, J. W. Drijfhout, R. Jordens, J. J. Onderwater, D. Verwoerd, A. A. Mulder, A. N. van der Heiden, D. Scheidegger, L. C. Oomen, T. H. Ottenhoff, A. Tulp, J. J. Neefjes, and F. Koning (1997). Mannose receptor-mediated uptake of antigens strongly enhances HLA class II-restricted antigen presentation by cultured dendritic cells. *Eur. J. Immunol.* **27**, 2426–35.

12. Engering, A. J., M. Cella, D. Fluitsma, M. Brockhaus, E. C. Hoefsmit, A. Lanzavecchia, and J. Pieters (1997). The mannose receptor functions as a high capacity and broad specificity antigen receptor in human dendritic cells. *Eur. J. Immunol.* **27**, 2417–25.

13. Inohara, N., M. Chamaillard, C. McDonald, and G. Nunez (2005). NOD-LRR proteins: role in host–microbial interactions and inflammatory disease. *Annu. Rev. Biochem.* **74**, 355–83.

14. Wagner, H. (2004). The immunobiology of the TLR9 subfamily. *Trends Immunol.* **25**, 381–6.

15. McGreal, E. P., J. L. Miller, and S. Gordon (2005). Ligand recognition by antigen-presenting cell C-type lectin receptors. *Curr. Opin. Immunol.* **17**, 18–24.

16. Kokkinopoulos, I., W. J. Jordan, and M. A. Ritter (2005). Toll-like receptor mRNA expression patterns in human dendritic cells and monocytes. *Mol. Immunol.* **42**, 957–68.

17. Portnoy, D. A. (2005). Manipulation of innate immunity by bacterial pathogens. *Curr. Opin. Immunol.* **17**, 25–8.

18. Kawasaki, K., R. K. Ernst, and S. I. Miller (2004). 3-O-deacylation of lipid A by PagL, a PhoP/PhoQ-regulated deacylase of *Salmonella typhimurium*, modulates signaling through Toll-like receptor 4. *J. Biol. Chem.* **279**, 20044–8.

19. Darveau, R. P., T. T. Pham, K. Lemley, R. A. Reife, B. W. Bainbridge, S. R. Coats, W. N. Howald, S. S. Way, and A. M. Hajjar (2004). Porphyromonas gingivalis lipopolysaccharide contains multiple lipid A

species that functionally interact with both Toll-like receptors 2 and 4. *Infect. Immun.* **72**, 5041–51.

20. Netea, M. G., J. W. Van Der Meer, and B. J. Kullberg (2004). Toll-like receptors as an escape mechanism from the host defense. *Trends Microbiol.* **12**, 484–8.

21. van der Kleij, D., E. Latz, J. F. Brouwers, Y. C. Kruize, M. Schmitz, E. A. Kurt-Jones, T. Espevik, E. C. de Jong, M. L. Kapsenberg, D. T. Golenbock, A. G. Tielens, and M. Yazdanbakhsh (2002). A novel host-parasite lipid cross-talk. Schistosomal lyso-phosphatidylserine activates toll-like receptor 2 and affects immune polarization. *J. Biol. Chem.* **277**, 48122–9.

22. Sing, A., D. Rost, N. Tvardovskaia, A. Roggenkamp, A. Wiedemann, C. J. Kirschning, M. Aepfelbacher, and J. Heesemann (2002). Yersinia V-antigen exploits toll-like receptor 2 and CD14 for interleukin 10-mediated immunosuppression. *J. Exp. Med.* **196**, 1017–24.

23. Geijtenbeek, T. B. H., R. Torensma, S. J. van Vliet, G. C. F. van Duijnhoven, G. J. Adema, Y. van Kooyk, and C. G. Figdor (2000). Identification of DC-SIGN, a novel dendritic cell-specific ICAM-3 receptor that supports primary immune responses. *Cell* **100**, 575–85.

24. Geijtenbeek, T. B., D. J. Krooshoop, D. A. Bleijs, S. J. van Vliet, G. C. van Duijnhoven, V. Grabovsky, R. Alon, C. G. Figdor, and Y. van Kooyk (2000). DC-SIGN-ICAM-2 interaction mediates dendritic cell trafficking. *Nat. Immunol.* **1**, 353–7.

25. Ariizumi, K., G. L. Shen, S. Shikano, S. Xu, R. Ritter, T. Kumamoto, D. Edelbaum, A. Morita, P. R. Bergstresser, and A. Takashima (2000). Identification of a novel, dendritic cell-associated molecule, dectin-1, by subtractive cDNA cloning. *J. Biol. Chem.* **275**, 20157–67.

26. Irjala, H., E. L. Johansson, R. Grenman, K. Alanen, M. Salmi, and S. Jalkanen (2001). Mannose receptor is a novel ligand for L-selectin and mediates lymphocyte binding to lymphatic endothelium. *J. Exp. Med.* **194**, 1033–42.

27. Stahl, P. D. (1992). The mannose receptor and other macrophage lectins. *Curr. Opin. Immunol.* **4**, 49–52.

28. Hawiger, D., K. Inaba, Y. Dorsett, M. Guo, K. Mahnke, M. Rivera, J. V. Ravetch, R. M. Steinman, and M. C. Nussenzweig (2001). Dendritic cells induce peripheral T cell unresponsiveness under steady state conditions in vivo. *J. Exp. Med.* **194**, 769–79.

29. Chieppa, M., G. Bianchi, A. Doni, A. Del Prete, M. Sironi, G. Laskarin, P. Monti, L. Piemonti, A. Biondi, A. Mantovani, M. Introna, and P. Allavena (2003). Cross-linking of the mannose receptor on monocyte-derived

dendritic cells activates an anti-inflammatory immunosuppressive program. *J. Immunol.* **171**, 4552−60.

30. Kwon, D. S., G. Gregorio, N. Bitton, W. A. Hendrickson, and D. R. Littman (2002). DC-SIGN-mediated internalization of HIV is required for trans-enhancement of T cell infection. *Immunity* **16**, 135−44.

31. Ludwig, I. S., A. N. Lekkerkerker, E. Depla, F. Bosman, R. J. Musters, S. Depraetere, Y. van Kooyk, and T. B. Geijtenbeek (2004). Hepatitis C virus targets DC-SIGN and L-SIGN to escape lysosomal degradation. *J. Virol.* **78**, 8322−32.

32. Geijtenbeek, T. B. H., D. S. Kwon, R. Torensma, S. J. van Vliet G. C. F. van Duijnhoven, J. Middel, I. L. Cornelissen, H. S. Nottet, V. N. KewalRamani, D. R. Littman, C. G. Figdor, and Y. van Kooyk (2000). DC-SIGN, a dendritic cell-specific HIV-1-binding protein that enhances trans-infection of T cells. *Cell* **100**, 587−97.

33. Lozach, P. Y., A. Amara, B. Bartosch, J. L. Virelizier, F. Arenzana-Seisdedos, F. L. Cosset, and R. Altmeyer (2004). C-type lectins L-SIGN and DC-SIGN capture and transmit infectious hepatitis C virus pseudotype particles. *J. Biol. Chem.* **279**, 32035−45.

34. Alvarez, C. P., F. Lasala, J. Carrillo, O. Muniz, A. L. Corbi, and R. Delgado (2002). C-type lectins DC-SIGN and L-SIGN mediate cellular entry by Ebola virus in cis and in trans. *J. Virol.* **76**, 6841−4.

35. Halary, F., A. Amara, H. Lortat-Jacob, M. Messerle, T. Delaunay, C. Houles, F. Fieschi, F. Arenzana-Seisdedos, J. F. Moreau, and J. Dechanet-Merville (2002). Human cytomegalovirus binding to DC-SIGN is required for dendritic cell infection and target cell trans-infection. *Immunity* **17**, 653−64.

36. Tassaneetrithep, B., T. H. Burgess, A. Granelli-Piperno, C. Trumpfheller, J. Finke, W. Sun, M. A. Eller, K. Pattanapanyasat, S. Sarasombath, D. L. Birx, R. M. Steinman, S. Schlesinger, and M. A. Marovich (2003). DC-SIGN (CD209) mediates dengue virus infection of human dendritic cells. *J. Exp. Med.* **197**, 823−9.

37. Appelmelk, B. J., I. van Die, S. J. van Vliet, C. M. Vandenbroucke-Grauls, T. B. Geijtenbeek, and Y. van Kooyk (2003). Cutting edge: carbohydrate profiling identifies new pathogens that interact with dendritic cell-specific ICAM-3-grabbing nonintegrin on dendritic cells. *J. Immunol.* **170**, 1635−9.

38. Van Die, I., S. J. van Vliet, A. Kwame Nyame, R. D. Cummings, C. M. Bank, B. Appelmelk, T. B. Geijtenbeek, and Y. van Kooyk (2003). The dendritic cell specific C-type lectin DC-SIGN is a receptor for *Schistosoma mansoni* egg antigens and recognizes the glycan antigen Lewis-x. *Glycobiology* **13**, 471−8.

39. Cambi, A., K. Gijzen, J. M. de Vries, R. Torensma, B. Joosten, G. J. Adema, M. G. Netea, B. J. Kullberg, L. Romani, and C. G. Figdor (2003). The C-type lectin DC-SIGN (CD209) is an antigen-uptake receptor for *Candida albicans* on dendritic cells. *Eur. J. Immunol.* **33**, 532–8.

40. Geijtenbeek, T. B., S. J. van Vliet, E. A. Koppel, M. Sanchcz-Hernandez, C. M. Vandenbroucke-Grauls, B. Appelmelk, and Y. van Kooyk (2003). Mycobacteria target DC-SIGN to suppress dendritic cell function. *J. Exp. Med.* **197**, 7–17.

41. Tailleux, L., O. Schwartz, J. L. Herrmann, E. Pivert, M. Jackson, A. Amara, L. Legres, D. Dreher, L. P. Nicod, J. C. Gluckman, P. H. Lagrange, B. Gicquel, and O. Neyrolles (2003). DC-SIGN is the major *Mycobacterium tuberculosis* receptor on human dendritic cells. *J. Exp. Med.* **197**, 121–7.

42. Baldari, C. T., A. Lanzavecchia, and J. L. Telford (2005). Immune subversion by *Helicobacter pylori*. *Trends Immunol.* **26**, 199–207.

43. Bergman, M. P., A. Engering, H. H. Smits, S. J. van Vliet, A. A. van Bodegraven, H. P. Wirth, M. L. Kapsenberg, C. M. Vandenbroucke-Grauls, Y. van Kooyk, and B. J. Appelmelk (2004). *Helicobacter pylori* modulates the T helper cell 1/T helper cell 2 balance through phase-variable interaction between lipopolysaccharide and DC-SIGN. *J. Exp. Med.* **200**, 979–90.

44. Appelmelk, B. J., M. A. Monteiro, S. L. Martin, A. P. Moran, and C. M. Vandenbroucke-Grauls (2000). Why *Helicobacter pylori* has Lewis antigens. *Trends Microbiol.* **8**, 565–70.

45. Smits, H. H., A. Engering, D. van der Kleij, E. C. de Jong, K. Schipper, T. van Capel, B. Zaat, M. Yazdanbakhsh, E. A. Wierenga, Y. Kooyk, and M. L. Kapsenberg (2005). Selective probiotic bacteria induce regulatory T cells by modulating dendritic cell function via DC-SIGN in vitro. *J. Allergy and Clin. Immunol.* **115**, 1260–7.

46. Geijtenbeek, T. B., P. C. Groot, M. A. Nolte, S. J. van Vliet, S. T. Gangaram-Panday, G. C. van Duijnhoven, G. Kraal, A. J. van Oosterhout, and Y. van Kooyk (2002). Marginal zone macrophages express a murine homologue of DC-SIGN that captures blood-borne antigens in vivo. *Blood* **100**, 2908–16.

47. Baribaud, F., S. Pohlmann, and R. W. Doms (2001). The role of DC-SIGN and DC-SIGNR in HIV and SIV attachment, infection, and transmission. *Virology* **286**, 1–6.

48. Taylor, P. R., G. D. Brown, J. Herre, D. L. Williams, J. A. Willment, and S. Gordon (2004). The role of SIGNR1 and the beta-glucan receptor (dectin-1) in the nonopsonic recognition of yeast by specific macrophages. *J. Immunol.* **172**, 1157–62.

49. Kang, Y. S., J. Y. Kim, S. A. Bruening, M. Pack, A. Charalambous, A. Pritsker, T. M. Moran, J. M. Loeffler, R. M. Steinman, and C. G. Park (2004). The C-type lectin SIGN-R1 mediates uptake of the capsular polysaccharide of *Streptococcus pneumoniae* in the marginal zone of mouse spleen. *Proc. Natl Acad. Sci. USA* **101**, 215–20.

50. Takahara, K., Y. Yashima, Y. Omatsu, H. Yoshida, Y. Kimura, Y. S. Kang, R. M. Steinman, C. G. Park, and K. Inaba (2004). Functional comparison of the mouse DC-SIGN, SIGNR1, SIGNR3 and Langerin, C-type lectins. *Int. Immunol.* **16**, 819–29.

51. Lanoue, A., M. R. Clatworthy, P. Smith, S. Green, M. J. Townsend, H. E. Jolin, K. G. Smith, P. G. Fallon, and A. N. McKenzie (2004). SIGN-R1 contributes to protection against lethal pneumococcal infection in mice. *J. Exp. Med.* **200**, 1383–93.

52. Nagaoka, K., K. Takahara, K. Tanaka, H. Yoshida, R. M. Steinman, S. I. Saitoh, S. Akashi-Takamura, K. Miyake, Y. S. Kang, C. G. Park, and K. Inaba (2005). Association of SIGNR1 with TLR4-MD-2 enhances signal transduction by recognition of LPS in Gram-negative bacteria. *Int. Immunol.* **17**, 827–36.

53. Brown, G. D. and S. Gordon (2001). Immune recognition. A new receptor for beta-glucans. *Nature* **413**, 36–7.

54. Underhill, D. M., A. Ozinsky, A. M. Hajjar, A. Stevens, C. B. Wilson, M. Bassetti, and A. Aderem (1999). The Toll-like receptor 2 is recruited to macrophage phagosomes and discriminates between pathogens. *Nature* **401**, 811–15.

55. Brown, G. D., J. Herre, D. L. Williams, J. A. Willment, A. S. Marshall, and S. Gordon (2003). Dectin-1 mediates the biological effects of beta-glucans. *J. Exp. Med.* **197**, 1119–24.

56. Gantner, B. N., R. M. Simmons, S. J. Canavera, S. Akira, and D. M. Underhill (2003). Collaborative induction of inflammatory responses by dectin-1 and Toll-like receptor 2. *J. Exp. Med.* **197**, 1107–17.

57. Herre, J., A. S. Marshall, E. Caron, A. D. Edwards, D. L. Williams, E. Schweighoffer, V. Tybulewicz, Reis E Sousa, S. Gordon, and G. D. Brown (2004). Dectin-1 uses novel mechanisms for yeast phagocytosis in macrophages. *Blood* **104**, 4038–45.

58. Rogers, N. C., E. C. Slack, A. D. Edwards, M. A. Nolte, O. Schulz, E. Schweighoffer, D. L. Williams, S. Gordon, V. L. Tybulewicz, G. D. Brown, and Reis E Sousa (2005). Syk-dependent cytokine induction by dectin-1 reveals a novel pattern recognition pathway for C type lectins. *Immunity* **22**, 507–17.

59. Gantner, B. N., R. M. Simmons, and D. M. Underhill (2005). Dectin-1 mediates macrophage recognition of *Candida albicans* yeast but not filaments. *EMBO J.* **24**, 1277–86.

60. d'Ostiani, C. F., G. Del Sero, A. Bacci, C. Montagnoli, A. Spreca, A. Mencacci, P. Ricciardi-Castagnoli, and L. Romani (2000). Dendritic cells discriminate between yeasts and hyphae of the fungus *Candida albicans*. Implications for initiation of T helper cell immunity in vitro and in vivo. *J. Exp. Med.* **191**, 1661–74.

61. Romani, L., C. Montagnoli, S. Bozza, K. Perruccio, A. Spreca, P. Allavena, S. Verbeek, R. A. Calderone, F. Bistoni, and P. Puccetti (2004). The exploitation of distinct recognition receptors in dendritic cells determines the full range of host immune relationships with *Candida albicans*. *Int. Immunol.* **16**, 149–61.

62. Huang, Q., D. Liu, P. Majewski, L. C. Schulte, J. M. Korn, R. A. Young, E. S. Lander, and N. Hacohen (2001). The plasticity of dendritic cell responses to pathogens and their components. *Science* **294**, 870–5.

63. Jeannin, P., B. Bottazzi, M. Sironi, A. Doni, M. Rusnati, M. Presta, V. Maina, G. Magistrelli, J. F. Haeuw, G. Hoeffel, N. Thieblemont, N. Corvaia, C. Garlanda, Y. Delneste, and A. Mantovani (2005). Complexity and complementarity of outer membrane protein A recognition by cellular and humoral innate immunity receptors. *Immunity* **22**, 551–60.

64. Pinhal-Enfield, G., M. Ramanathan, G. Hasko, S. N. Vogel, A. L. Salzman, G. J. Boons, and S. J. Leibovich (2003). An angiogenic switch in macrophages involving synergy between Toll-like receptors 2, 4, 7, and 9 and adenosine A(2A) receptors. *Am. J. Pathol.* **163**, 711–21.

65. Mantovani A., S. Sozzani, M. Locati, P. Allavena, and A. Sica (2002). Macrophage polarization: tumor-associated macrophages as a paradigm for polarized M2 mononuclear phagocytes. *Trends Immunol.* **23**, 549–55.

66. Mosser, D. M. (2003). The many faces of macrophage activation. *J. Leukoc. Biol.* **73**, 209–12.

67. Raes, G., L. Brys, B. K. Dahal, J. Brandt, J. Grooten, F. Brombacher, G. Vanham, W. Noel, P. Bogaert, T. Boonefaes, A. Kindt, B. R. Van den, P. J. Leenen, P. De Baetselier, and G. H. Ghassabeh (2005). Macrophage galactose-type C-type lectins as novel markers for alternatively activated macrophages elicited by parasitic infections and allergic airway inflammation. *J. Leukoc. Biol.* **77**, 321–7.

68. Higashi, N., K. Fujioka, K. Denda-Nagai, S. Hashimoto, S. Nagai, T. Sato, Y. Fujita, A. Morikawa, M. Tsuji, M. Miyata-Takeuchi, Y. Sano, N. Suzuki, K. Yamamoto, K. Matsushima, and T. Irimura (2002). The macrophage

C-type lectin specific for galactose/N-acetylgalactosamine is an endocytic receptor expressed on monocyte-derived immature dendritic cells. *J. Biol. Chem.* **277**, 20686–93.

69. van Vliet, S. J., E. van Liempt, E. Saeland, C. A. Aarnoudse, B. Appelmelk, T. Irimura, T. B. Geijtenbeek, O. Blixt, R. Alvarez, I. van Die, and Y. van Kooyk (2005). Carbohydrate profiling reveals a distinctive role for the C-type lectin MGL in the recognition of helminth parasites and tumor antigens by dendritic cells. *Int. Immunol.* **17**, 661–9.

A. ENGERING, S. J. VAN VLIET, E. A. KOPPEL *ET AL.*

CHAPTER 10

Suppression of immune responses by bacteria and their products through dendritic cell modulation and regulatory T cell induction

Miriam T. Brady, Peter McGuirk and Kingston H. G. Mills
Trinity College, Dublin

10.1 INTRODUCTION

Infection with pathogenic bacteria can result in acute or chronic disease, which can be life threatening, especially in young, elderly or other immunocompromised individuals. Humans are also infected with a wide range of commensal bacteria, as part of our normal gut flora, and the immune system must be capable of controlling immune responses against these beneficial bacteria, while at the same time generating effector immune responses against pathogenic micro-organisms. In addition, pathogenic bacteria have evolved strategies for delaying or preventing their elimination by evading or subverting protective immune responses of the host.

10.1.1 Innate immunity to bacteria

The initial inflammatory response to pathogenic bacteria involves the release of cytokines and chemokines and the recruitment of neutrophils, monocytes, dendritic cells (DCs) and lymphocytes to the site of infection. Tissue macrophages and neutrophils quickly phagocytose and attempt to kill the bacteria. Macrophages and DCs are activated through binding of conserved, secreted or cell surface bacterial products to pathogen recognition receptors (PRR). This leads to activation of immune response genes, including those coding for inflammatory cytokines, chemokines and co-stimulatory molecules expressed on the surface of DCs and macrophages, that are involved in antigen presentation (Janeway and Medzhitov, 2002).

Bacteria are phagocytosed by neutrophils and macrophages and this is facilitated through activation of the alternative complement pathway by bacterial cell wall components, resulting in the production of C3b, which together with antibodies help to opsonize the bacteria. Once inside the phagocytic cells, the bacteria-containing phagosome undergoes a series of fusions with endosomes and lysosomes to form a phagolysosome where the bacteria are killed in a variety of ways, including reactive oxygen and nitrogen-dependent mechanisms. However, bacteria can inhibit killing by preventing phagosome–lysosome fusion or by inducing lysosomal discharge into the cytoplasm. Induction of apoptosis in neutrophils macrophages and DCs, critical cells in anti-bacterial host defence and in activating adaptive immunity, is an alternative immune evasion strategy evolved by a number of bacterial species to subvert protective immune responses of the host.

10.1.2 Adaptive immunity to bacteria

Following uptake of the bacteria by a macrophage or DC, these professional antigen presenting cells (APC) process and present peptides of the bacterial antigens to T cells, in association with major histocompatibility complex (MHC) molecules on their cell surface. All bacteria, whether they survive intracellularly or extracellularly in the host, can activate MHC class II restricted $CD4^+$ T helper (Th) cells. In addition, certain intracellular bacteria can also activate MHC class I-restricted $CD8^+$ cytotoxic T lymphocytes (CTL). Induction of class II-restricted T cells requires processing by an exogenous route. Class I restricted CTL kill host cells infected with intracellular bacteria, whereas the main function of class II-restricted $CD4^+$ T cells is to release cytokines that activate phagocytosis and killing of bacteria by macrophages and to provide helper function for antibody production. $CD4^+$ T cells can be divided into a number of functionally distinct subtypes discriminated on the basis of cytokine secretion. Th1 cells secrete interferon-γ (IFN-γ) and tumor necrosis factor-β (TNF-β) and activate phagocytosis and killing by macrophages and provide help for the production of opsonizing (murine IgG2a) antibodies. More recently a distinct population of effector T cells that secrete IL-17 and promote inflammatory reactions in autoimmune diseases also appear to play a protective role in immunity to certain bacteria (Happel et al., 2003). Th2 cells secrete IL-4, IL-5, IL-6, IL-9, IL-10 and IL-13 and are considered to be the main helper cells, especially for IgG1, IgA and IgE. Finally, T cells that secrete IL-10 and/or TGF-β, termed regulatory

T (Treg) cells, that suppress immune responses mediated by innate immune cells and other T cells, can be induced by infection (Mills, 2004).

The control of intracellular bacteria is dependent on the induction of cell-mediated immunity, however, humoral immunity plays a major role in protection against extracellular bacteria. In individuals that have recovered from self-limiting bacterial infections, antibodies play a major role in preventing re-infection. Furthermore successful bacterial vaccines confer protection by the generation of circulating IgG or memory B cells, which produce an anamnestic antibody response following re-exposure to the bacteria. However, the generation of antibody responses is also dependent on priming of helper T cells, so both T and B cells are critical in the primary as well as secondary response to bacterial infection. Ig class switching to IgG2a in the mouse is promoted by IFN-γ-secreting Th1 cells and this antibody subclass is involved in opsonization of bacteria and also stimulates complement components C3b and iC3b, which bind complement receptors and further promote phagocytosis. Extracellular bacteria often bind to epithelial cells in the respiratory or gastrointestinal (GI) tract via adhesins or pili, and antibodies against these bacterial virulence factors can help to prevent colonization. Antibodies of the murine IgG1 subclass can neutralize bacterial toxins and prevent their binding to host target cells and thereby reduce the severity of diseases caused by toxin-producing bacteria. IgA antibodies induced following bacterial infection of the respiratory tract, GI tract or other mucosal tissues, function to limit the infection to mucosal surfaces.

In addition to CD4$^+$ and CD8$^+$ T cells, unconventional T cells, including $\gamma\delta$ T cells which recognize phospholigands and CD1-restricted $\alpha\beta$ T cells, or T cells that express natural killer (NK) markers, termed NKT cells, that recognize glycolipids, also play a role in immunity to intracellular bacteria (Schaible and Kaufmann, 2000). Although it takes days rather than hours before the adaptive immune response is effective, unlike the innate immune system, it is able to recall previous encounters with antigen, through the activation of memory T and B cells. Therefore a second or subsequent infection with the bacteria is dealt with more effectively.

10.2 T CELL ANERGY OR SUPPRESSION INDUCED BY BACTERIA AND THEIR PRODUCTS

T cells play a major role in clearance of a primary infection with pathogenic bacteria and are fundamental in the memory response induced by previous infection or vaccination. However in many bacterial infections,

T cell responses either fail to develop, are suppressed, anergic or are skewed to an inappropriate subtype required for bacterial elimination. T cell activation is dependent on three signals. Signal one is provided through engagement of the T cell receptor (TCR) with processed antigenic peptide associated with a MHC molecule expressed on the surface of an APC. Signal two is provided by interaction of CD28 on the T cell surface with co-stimulatory molecules, CD80 or CD86 expressed on the surface of activated APC. In the absence of the second signal the T cells become anergic and are no longer able to divide or respond to antigen. A third signal which determines the polarization of the T cell response to distinct subtypes is provided by regulatory cytokines, such as IL-4, IL-6, IL-10, IL-12, IL-23 and IL-27, secreted primarily by cells of the innate immune system.

Many bacteria have evolved strategies to evade adaptive immunity by suppressing various steps in T cell activation or polarization, or by inducing a state of anergy, whereby T cells are unable to respond to the foreign pathogen. Interference with any aspect of antigen uptake processing or presentation by the APC, or signaling in the T cell can inhibit T cell activation. Furthermore, since activation of distinct T cell subtypes are regulated by the reciprocal subtype, factors that strongly promote the induction of one subtype may inhibit the activation of another. The Th1/Th2 paradigm has provided a simple model to explain the persistence of certain bacterial infections through pathogen induction of the reciprocal subtype; IFN-γ secreted by Th1 cells can suppress Th2 responses, whereas IL-4 and IL-10 secreted by Th2 cells can suppress Th1 responses. However, recent advances in our understanding of the role of Treg cells has complicated this model and it now appears that natural and inducible Treg cells may play a critical role in controlling both Th1 and Th2 cells and may also be responsible for anergy or immunosuppression observed during certain infections (Mills, 2004).

Salmonella typhimurium infection of mice results in profound immunosuppression, with inhibition of T and B cell proliferation and IL-2 production in response to foreign antigen and mitogens (al-Ramadi *et al.*, 1991a,b). Suppression with *S. typhimurium* was linked with soluble factors released by monocytes/macrophages and was reversed by the addition of IL-4. Nitric oxide (NO) was later shown to be involved in the immunosuppression as treatment of mice with the NO inhibitor, amino-guanidine hemisulfate, blocks the suppressive effect of *Salmonella* infection on T and B cell responses (MacFarlane *et al.*, 1999). However the NO inhibitor also blocked the influx of neutrophils and macrophages into the spleen and enhanced bacterial load, resulting in higher mortality of the mice, suggesting that NO is involved in host defence. *Helicobacter pylori* infection is also associated with immunosuppression and in vitro studies demonstrated that *H. pylori* suppressed proliferative responses of human PBMC to mitogens and antigens. A soluble cytoplasmic fraction of *H. pylori* was found to mediate the suppression by acting on monocytes and directly on the T cells (Knipp *et al.*, 1994).

In certain bacterial infections where immunosuppression has been reported, the bacterial virulence factor involved has been identified. The YopH protein of *Yersinia pseudotuberculosis*, which has tyrosine phosphatase activity, has been shown to suppress immune responses by interfering with T and B cell antigen-receptor activation (Yao *et al.*, 1999). The OspA protein of *Borrelia burgdorferi* inhibits proliferative responses of human PBMC to mitogens (Chiao *et al.*, 2000). Furthermore *B. burgdorferi* infection of disease susceptible (C3H/HeJ) and resistant (BALB/c) mice results in impaired proliferation, and IL-2 and IL-4 production to mitogens (de Souza *et al.*, 1993). More recently it has been suggested that Th1 responses are suppressed while Th2 responses are enhanced; *B. burgdorferi* transmission by *Ixodes scapularis* suppressed IL-2 and IFN-γ and enhanced IL-4 production in mice (Zeidner *et al.*, 1997). Similarly, the inhibitory effect of *Escherichia coli* LT on T cell responses was shown to be specific for Th1 cells and was mediated by the effect of ADP-ribosyl transferase enzyme activity on APC as well as on T cells (Ryan *et al.*, 2000). However, recent studies with the related AB type toxin, cholera toxin (CT), have suggested that that these toxins induce suppressive Treg cells, as well as Th2 cells (Lavelle *et al.*, 2003). It has also been reported that *E. coli* heat labile enterotoxin (LT)-treated epithelial cells release soluble factors, probably prostaglandins, that inhibit T cell proliferation in vitro (Lopes *et al.*, 2000).

Patients with lepromatous leprosy are highly immunosuppressed; their T cells do not respond to *Mycobacterium leprae* antigens and are anergized to unrelated antigens (Mehra *et al.*, 1984; Salgame *et al.*, 1984). The immunosuppression has been linked to macrophages, CD8$^+$ T cells, inappropriate Th1/Th2 induction or IL-10 production. Lipoarabinomannan (LAM)-B protein from *M. leprae* can suppress proliferative responses of PBMC from lepromatous leprosy patients, tuberculoid leprosy patients and from normal individuals (Kaplan *et al.*, 1987). Lipoglycans from *M. leprae* have also been shown to induce immune suppression in mice, including inhibition of delayed type hypersensitivity responses and proliferation of lymph node cells to mitogens (Moura *et al.*, 1997). In contrast to patients with lepromatous leprosy, where T cell responses to many antigens are profoundly suppressed and the bacteria persist, patients with tuberculoid leprosy have potent cellular immune responses and control the infection (Modlin, 1994). The patients with lepromatous leprosy mount Th2 responses, which are not protective, whereas in the tuberculoid leprosy patients Th1 cells are dominant and IFN-γ secreted by these cells activate infected macrophages to kill the bacteria.

Inappropriate Th2 cell induction and/or suppression of Th1 cells have also been shown to result in increased susceptibility to bacterial infections. Progression of *B. burgdorferi* infection following transmission by *I. scapularis*, in disease-susceptible C3H/HeJ mice is associated with the development of Th2 responses, but not in disease-resistant BALB/c mice (Zeidner *et al.*, 1997). IFN-γ secreted by NK cells during the acute stages of infection, and by Th1 cells later in infection, have been shown to play a crucial role in the clearance of *Bordetella pertussis* from the respiratory tract; mice depleted of NK cells or with defective IFN-γ receptors develop disseminating lethal infections (Byrne *et al.*, 2004; Mahon *et al.*, 1997). However, *B. pertussis* infection of immunocompetent individuals persists for several weeks or months. Furthermore, studies in a mouse model have shown that Th1 responses are suppressed in the lungs and draining lymph nodes during acute infection and this has been linked to the induction of innate anti-inflammatory cytokines and Treg cells by *B. pertussis* virulence factors (McGuirk *et al.*, 2002; McGuirk and Mills, 2000; Ross *et al.*, 2004).

10.3 MODULATION OF DC CYTOKINE PRODUCTION BY BACTERIA AND BACTERIAL MOLECULES

Dendritic cells utilize TLRs and other PRRs, including C-type lectins, to recognize characteristic conserved molecular patterns in microbial cell-wall components. Binding of PRRs on DCs with these pathogen associated molecular patterns (PAMPs) and consequent antigen capture, results in a cascade of events ultimately resulting in the activation of an appropriate immune response and elimination of the pathogen.

Following antigen capture in peripheral mucosal tissues, DCs migrate to secondary lymphoid organs for efficient antigen presentation. Concomitantly DCs undergo considerable changes including the activation of immune response genes, inflammatory cytokine and chemokine production, and provision of appropriate co-receptor signaling required for effective T-cell stimulation (Janeway and Medzhitov, 2002). Different microbial stimuli confer distinctive signals to the DCs via PRRs, modulating the

DC response and the subsequent differentiation of naive T-cells. In addition, pro-inflammatory cytokines and chemokines such as TNF-α, IL-1, IL-12 and MIPs promote the infiltration of leucocytes to the site of infection following bacterial challenge. These cytokines also play a crucial role in enhancing the bactericidal activity of phagocytes and in directing the immune response through the modulation of DCs. In this way the induction of protective immunity by DCs is highly dependent on the stimulus received and cytokines induced, and is therefore susceptible to manipulation by pathogens, which have evolved mechanisms to subvert DC function and escape immune surveillance. This immune regulation is a universal concept that includes suppression, diversion and conversion of the immune response to the benefit of the pathogen.

10.3.1 Suppression of IL-12 and TNF-α

A common strategy utilized by bacteria to modulate host responses is to suppress the immediate inflammatory response, normally associated with the production of proinflammatory cytokines, such as IL-12 and TNF-α. This is achieved either through the production of IL-10, TGF-β, or interference with signaling pathways.

Immunosuppression is an inherent complication of mycobacterial infections. Several studies have shown that *M. tuberculosis* or *M. bovis* bacillus Calmette-Guerin (BCG) target the C-type lectin DC-SIGN (DC-specific intercellular adhesion molecule-grabbing nonintegrin) to infect DCs and inhibit their immunostimulatory function (Geijtenbeek et al., 2003; van Kooyk and Geijtenbeek, 2003). This is thought to occur through the interaction of the mycobacterial mannosylated LAM with DC-SIGN, which can inhibit IL-12 production (Gagliardi et al., 2005; Nigou et al., 2001), and prevent DC maturation (Geijtenbeek et al., 2003) (see also Chapter 9). However, recently it has been shown that BCG-induced impairment of IL-12 from DCs cannot be attributed to the sole engagement of this receptor, as it occurs irrespective of DC-SIGN expression (Gagliardi et al., 2005).

Filamentous hemagglutinin (FHA) is an adhesin molecule and virulence factor from *B. pertussis*, which binds to the β2 integrin CR3 (Ishibashi et al., 1994). This bacterial molecule which interacts directly with DCs and macrophages also has immunomodulatory properties and has been shown to inhibit IL-12 production, partly through induction of IL-10 (McGuirk et al., 2002; McGuirk and Mills, 2000). In addition, FHA is capable of inhibiting both the Th1 response to influenza virus when administered simultaneously in the respiratory tract (McGuirk *et al.*, 2000), and LPS-driven IL-12 and IFN-γ production in vivo, in a murine model of septic shock (McGuirk and Mills, 2000). The ability of FHA to suppress IL-12 production is particularly relevant given that during acute infection, antigen-specific Th1 responses in the lung and its draining lymph nodes are also severely suppressed (McGuirk and Mills, 2000).

A number of bacterial toxins can also inhibit inflammatory responses by cells of the innate immune system. They include the AB-type toxins CT from *Vibrio cholerae* (Braun *et al.*, 1999), adenylate cyclase toxin (CyaA) from *B. pertussis* (Mills, 2001) and *E. coli* LT (Ryan *et al.*, 2000). CT has the ability to prevent the production of bioactive IL-12 and expression of the IL-12 receptor β1 and β2 chains on human monocytes and DCs (Braun *et al.*, 1999), and can inhibit IL-12-mediated experimental colitis (Boirivant *et al.*, 2001). Similarly, CT inhibits LPS or CD40L-induced IL-12 and TNF-α and chemokine production by human DC (Gagliardi *et al.*, 2000). LT, and the partially toxic mutant LTR72 suppress production of IL-12 and TNF-α and the inflammatory chemokines MIP-1α and MIP-1β (Ryan *et al.*, 2000). Another LT mutant, LTK63, which is devoid of ADP-ribosylating activity, does not possess all the inhibitory properties of the wild-type or partially toxic mutant, suggesting that enzyme activity is in part responsible for suppression. The proinflammatory cytokine inhibition by both CT and LT has subsequently been shown to be mediated through a cAMP-dependent mechanism (Bagley *et al.*, 2002a; Ryan *et al.*, 2000). CyaA from *B. pertussis* can also mediate anti-inflammatory effects through cAMP, by inhibiting IL-12 and TNF-α production by human monocyte-derived DCs (Bagley *et al.*, 2002b), and has also been shown to inhibit LPS-induced IL-12 and TNF-α production from murine DCs (Ross *et al.*, 2004).

The major virulence factors from *Bacillus anthracis*, the causative agent of anthrax, are the toxins lethal toxin and edema toxin and are both inhibitory. Prestimulation of murine DCs with purified lethal toxin severely impairs IL-12, TNF-α, IL-1β and IL-6 production after subsequent stimulation with LPS, and impairs their ability to prime allogeneic CD4[+] T-cells (Agrawal *et al.*, 2003). Edema toxin, another ADP-ribosylating toxin, which exerts anti-inflammatory effects by increasing intracellular cAMP levels (Hoover *et al.*, 1994), has recently been shown to cooperate with lethal toxin to impair cytokine secretion during infection of DCs (Tournier *et al.*, 2005). Murine bone marrow-derived DC exhibit very different cytokine secretion patterns when infected with *B. anthracis* strains secreting lethal toxin or edema toxin or both, and when infected with a lethal toxin/edema toxin negative strain.

Edema toxin inhibits IL-12p70 and TNF-α secretion, whereas lethal toxin inhibits IL-10 and TNF-α production. Simultaneous secretion of both lethal toxin and edema toxin have a cumulative effect on the inhibition of TNF-α (Tournier *et al.*, 2005). Lethal toxin and edema toxin display dominant effects on IL-12p70 and IL-10, respectively, and therefore during infection, co-function to impair DC cytokine secretion.

Pasteurella multocida toxin (PMT), which is a major virulence factor in progressive atopic rhinitis in wild and domestic animals has also recently been demonstrated to suppress IL-12 production in human monocyte-derived DCs, and inhibit the mucosal adjuvant effects of CT in mice (Bagley *et al.*, 2005). TNF-α production is suppressed at the site of infection and in the Peyer's patches of mice infected with *Yersinia enterocolitica* (Beuscher *et al.*, 1995). Infection of DCs with wild-type *Y. enterocolitica* also suppresses the release of TNF-α, IL-10 and IL-12, while infection with plasmid-cured *Y. enterocolitica* or with a YopP-deficient mutant resulted in the production of these cytokines (Erfurth *et al.*, 2004).

Probiotics, which are defined as microbial organisms beneficial to the host following ingestion, are now widely accepted as having immuno-modulatory properties and playing an important role in the maintenance of homeostasis. In a recent study, it has been demonstrated that various *Lactobacilli* species can exert differing and opposing effects on DC activation (Christensen *et al.*, 2002). DCs treated with *L. reuteri* secrete high levels of IL-6 and IL-10 in contrast to *L. casei*-treated DCs which preferentially induce IL-12 production (Christensen *et al.*, 2002). The anti-inflammatory effect of some *Lactobacilli* has been most efficiently demon-strated in the treatment of colitis. A higher proportion of *Lactobacillus* species that inhibit TNF-α can be recovered from mice without colitis compared to mice with microbiota-dependent colitis (IL-10-deficient) (Pena *et al.*, 2004). Furthermore, treatment of *H. hepaticus*-induced colitis with *L. reuteri* and *L. paracasei*, lowers proinflammatory colonic cytokine (IL-12 and TNF-α) levels, and reduces intestinal inflammation (Pena *et al.*, 2005). *L. reuteri* and *L. rhamnosus* have also shown to be beneficial in the management of atopic dermatitis in children (Rosenfeldt *et al.*, 2003).

10.3.2 Induction of IL-10 and TGF-β

The anti-inflammatory cytokine, IL-10, which has the ability to inhibit responses of T-cells and macrophages and other cell types, plays a key role in immune regulation by microbial pathogens. Its principal function appears to be to limit and ultimately terminate inflammatory responses

(Moore *et al.*, 2001), but has been shown to be exploited by pathogens to delay their elimination from the host. These effects are primarily mediated via the modulation of APC function (McBride *et al.*, 2002; Redpath *et al.*, 2001). A number of bacteria including *B. pertussis* (McGuirk *et al.*, 2002), *Mycobacterium* spp. (Redpath *et al.*, 2001), *Yersinia* spp. (Erfurth *et al.*, 2004), *Listeria* spp. (Brzoza *et al.*, 2004) and *Lactobacilli* (Smits *et al.*, 2005), have been shown to induce IL-10 production from macrophages or DCs.

The perturbation of cytokine networks is increasingly recognized as a pathogenicity mechanism exploited by a number of bacteria. However, the bacterial molecule(s) responsible has only been identified in a limited number of cases. FHA, an adhesin from *B. pertussis* which binds CR3 (Ishibashi *et al.*, 1994), induces IL-10 production from DC (McGuirk *et al.*, 2002; McGuirk and Mills, 2000). *Y. enterocolitica* rLcrV (virulence associated V antigen) or infection with *Y. enterocolitica* can suppress TNF-α production in mice (Beuscher *et al.*, 1995; Sing *et al.*, 2002). This process has been attributed to IL-10 induction by the bacteria, since mice lacking IL-10 have significantly increased survival following infection, have lower bacterial numbers, and a distinctive absence of TNF-α suppression (Sing *et al.*, 2002). Furthermore, it has been suggested that induction of IL-10 from DCs after infection with *Y. enterocolitica* may contribute to the reduction in T-cell proliferation observed in vitro (Erfurth *et al.*, 2004).

Immunosuppression associated with mycobacteria and the suppressed type 1 responses of *M. tuberculosis* infected individuals has been attributed to the production of IL-10 and TGF-β production by mononuclear cells, including macrophages and DCs. These inhibitory cytokines have been implicated in the hyporesponsiveness observed in the lungs of patients with active pulmonary TB, providing a microenvironment wherein immune cells become refractory to appropriate activating signals (Bonecini-Almeida *et al.*, 2004). LAM, a cell wall component of *M. tuberculosis*, induces IL-10 production from human mononuclear cells and DCs (Barnes *et al.*, 1992; Chieppa *et al.*, 2003), and TGF-β from human monocytes (Dahl *et al.*, 1996), and this has also been provided as an explanation for the immunosuppressive effects of the bacteria on antigen-induced T-cell proliferation. LAMs from *M. tuberculosis* have also been shown to inhibit LPS-induced IL-12p70 and p40 production from human monocyte-derived DCs, an effect which is dependent on the presence of mannooligosaccharide caps (Nigou *et al.*, 2001). LAMs may therefore act as virulence factors for the bacteria, inducing anti-inflammatory cytokines and preventing the induction of IL-12, which is critical for the generation of Th1 responses required to effectively eliminate *M. tuberculosis*. *M. tuberculosis* secretory antigen can down-regulate Th1 responses to mycobacteria by differentially modulating the cytokine profiles of DCs. Restoration of the attenuated Th1 response can be achieved by blocking IL-10 or TGF-β with monoclonal antibodies, indicating a pivotal role for these cytokines in suppression of immune responses to mycobacteria (Balkhi *et al.*, 2004; Natarajan *et al.*, 2003). Similarly, suppression of IFN-γ production by purified protein derivative (PPD) from *M. tuberculosis* is abrogated following neutralization of IL-10 and TGF-β (Othieno *et al.*, 1999).

Studies in IL-10 knockout (IL-10$^{-/-}$) mice have provided more definitive evidence of an important role for endogenous or bacteria-stimulated IL-10 production on the course of bacterial infections. In certain cases infection is less severe in the absence of IL-10, confirming that this cytokine may subvert protective immunity by suppressing innate and adaptive immune responses. For example, lack of IL-10 during *M. bovis* BCG infection leads to accelerated clearance of bacilli, and enhanced plasma concentrations of the inflammatory cytokines IL-12 and TNF-α (Jacobs *et al.*, 2000; Jacobs *et al.*, 2002). Furthermore, in the absence of IL-10, BCG-infected DCs are more effective at trafficking to lymph nodes, produce significantly more IL-12, and enhance IFN-γ production in response to mycobacterial antigens compared to wildtype DCs (Demangel *et al.*, 2002). IL-10$^{-/-}$ mice are highly resistant to *Yersinia* infection and lack the IL-10-mediated TNF-α suppression observed in susceptible animals (Sing *et al.*, 2002). However, in other infections the role of IL-10 in terminating potentially adverse effects of inflammatory responses has been highlighted, by the observation that severity of disease is exacerbated in absence of this cytokine. In cerebral infection with *Listeria monocytogenes*, IL-10$^{-/-}$ mice succumb to primary and secondary infection, and recruit significantly more inflammatory cells to the brain. Furthermore, inflammatory cytokine production such as IL-1β, TNF-α and IL-12, and severity of brain lesions is enhanced in IL-10$^{-/-}$ mice (Deckert *et al.*, 2001). Despite this prominent hyperinflammation, intracerebral bacterial load is not reduced in IL-10$^{-/-}$ mice compared with wildtype mice, suggesting that IL-10 in this case plays a more significant role in damage limitation rather than protective immunity. However, TGF-β has been shown to protect against lethal infection and inhibit the production of inflammatory cytokines during listeriosis (Nakane *et al.*, 1996).

There are also examples of bacterial infection in IL-10$^{-/-}$ mice where disease pathology is enhanced, or where bacterial clearance is accelerated.

Colonization of the gastric mucosa by *Helicobacter pylori* is significantly reduced in IL-10$^{-/-}$ mice, but infiltration of inflammatory mediators, and severity of chronic active gastritis is enhanced compared to WT mice (Chen *et al.*, 2001). Similarly, endogenous IL-10 impairs bacterial clearance during *E. coli* peritonitis. However, despite lower bacterial numbers, IL-10$^{-/-}$ mice had higher concentrations of inflammatory cytokines and demonstrated more severe organ damage (Sewnath *et al.*, 2001). Hence, although IL-10 may enhance dissemination of the bacteria, it protects mice from lethality by attenuating immune-mediated pathology. IL-10$^{-/-}$ mice infected with *Helicobacter hepaticus* develop chronic colitis, with a critical role for IL-12 in both the induction and maintenance of the inflammatory process following infection (Kullberg *et al.*, 2001). In wild-type mice, infection with *H. hepaticus* induces Treg cells that prevent bacteria-induced colitis (Kullberg *et al.*, 2002). However, in the absence of IL-10 this protective mechanism fails, and mice develop chronic disease.

10.4 INFLUENCE OF BACTERIA AND BACTERIAL MOLECULES ON DC MATURATION

Dendritic cells are pivotal in the activation of naive T-cells and for the initiation of the primary immune response. The factors that determine DC function depend on the nature of the pathogen, the maturation status induced and the local environment. The process of maturation involves the continuous transition from an immature DCs residing in peripheral tissue and sampling its environment, to the mature DCs in the secondary lymphatic organs. These mature DCs are fully equipped for the efficient stimulation of Ag-specific T-cells.

Steady-state DCs are immature, have high phagocytic activity, express low-levels of costimulatory molecules and have low antigen processing capacity (Banchereau *et al.*, 2000). Microbial-induced maturation signals, such as LPS, TNF-α, CD40 ligand and antigen interactions instigate the maturation process. Depending on the initial signals or stimuli provided by the pathogen, various signaling pathways are triggered within the DCs, which then undergo considerable modification, such as induction of costimulatory molecules, cytokine production, migratory properties and endocytic activity or morphology. Additional DC stimulation is achieved and modulated by proinflammatory cytokines such as TNF, IL-1β and prostaglandin E2, or by a variety of non-inflammatory and pathogen unrelated factors like histamine and ATP. These supplementary stimuli influence the nature of a maturational process and provide DCs with different capacities for T cell effector subset priming (Kalinski *et al.*, 1999; Lanzavecchia and Sallusto, 2001). The upregulation of MHC and costimulatory molecules enhances the antigen presenting capacities of DCs and their T-cell stimulatory potential. This complex process culminates in the arrival of activated DCs in the T cell zone of lymph nodes where they interact with T cells. Therefore, the bacterial antigens that are able to induce maturation of DCs, and the extent of this maturation, help to define the character of primary immune responses against the pathogen and thus have an important role in determining the course of an infection (Reis e Sousa, 2004).

In addition to the modulation of DC cytokine production (discussed above), bacteria or their various virulence factors can also modulate MHC and co-stimulatory molecule expression on DCs. The AB-type toxins CT and LT induce maturation of human monocyte-derived DCs, with upregulation of CD80, CD86, CD83 and HLA-DR (Bagley *et al.*, 2002a). This maturation process is at least in part dependent on elevation of intracellular cAMP levels, which the toxins achieve through the constitutive activation of adenylate cyclase. Enzymatically inactive toxins such as CTK63 and LTK63 are unable to upregulate cAMP levels in the cell, and thus are less efficient in activating DCs. In addition, phenotypic maturity as determined by enhanced surface marker expression, correlates with an increased ability to stimulate proliferation of allogeneic T-cells. However, although the AB toxins mature DCs with regard to phenotype and functionality, they also profoundly inhibit IL-12 and TNF-α production (Bagley *et al.*, 2002a). CT has been shown to mature DCs which prime Th2 cells (Anjuere *et al.*, 2004; Gagliardi *et al.*, 2000), which may be due to its capacity to suppress the Th1-driving cytokine IL-12. CT also has the ability to promote the induction of Treg cells against bystander antigens, by modulating DC activation (Lavelle *et al.*, 2003). Stimulation of DC with CT enhanced expression of CD80, CD86, but downregulated CD40 expression, which is known to be important for induction of IL-12 and Th1 responses. Furthermore, adoptive transfer of DCs pulsed with antigen in the presence of CT, primed antigen-specific T-cells that produced IL-10, low levels of IL-4 and IL-5, and almost undetectable IFN-γ, a Tr1 type cell phenotype.

Bacillus anthracis, the causative agent of anthrax, has the ability to disrupt host immune responses by specifically targeting DCs. One of its critical virulence factors is anthrax lethal toxin, which not only modulates DC cytokine production but has a profound effect on the functional capacity of DCs (Agrawal *et al.*, 2003; Tournier *et al.*, 2005).

Upregulation of CD40, CD80 and CD86 on DCs is severely impaired following exposure to lethal toxin and these DCs fail to efficiently prime allogeneic CD4$^+$ T-cells and are refractory to further stimulus by the potent mitogen PMA and ionomycin (Agrawal *et al.*, 2003). The mechanism involved was shown to involve impaired phosphorylation of p38 and ERK1/ERK2. DC function is also perturbed during infection with *Y. enterocolitica*. Bacteria invade DC at the onset of infection and down-regulate MHC class II and CD80 molecules, which correlates with a decreased T-cell activation capacity (Schoppet *et al.*, 2000). The decrease in immunostimulatory ability is a transient effect, but may be sufficient to facilitate bacterial persistence in susceptible individuals. Similar effects have been observed in murine DCs (Erfurth *et al.*, 2004).

Coxiella burnetii, the etiological agent of the zoonotic disease Q fever, can evade immune clearance and result in a persistent infection. This pathogen has been shown to evade the protective response through manipulation of DC maturation. A virulent strain can infect human DCs but does not result in their maturation, demonstrated by the lack of inflammatory cytokine production, and upregulation of surface markers CD80, CD86, CD40, CD83 and HLA-DR. In contrast, an avirulent strain which possesses a severely truncated LPS, effectively induces DC maturation (Shannon *et al.*, 2005). The lack of DC maturation by the virulent strain results in an immune response that impedes clearance of the bacteria and allows persistence. The mechanism proposed is that the intact LPS of the virulent strain "masks" TLR ligands from recognition by DCs, preventing DC maturation and induction of the appropriate immune response for bacterial clearance.

10.5 TREG CELLS IN BACTERIAL INFECTIONS

10.5.1 Natural and inducible regulatory T cells

A population of CD4$^+$ T cells, that secrete high levels of IL-10 and/or TGF-β, but low or undetectable IL-4 and IFN-γ, termed inducible Treg cells have recently been described during infection with certain pathogens, including bacteria (McGuirk *et al.*, 2002). Treg cells have suppressive function and are induced by the host to control immune responses during infection, especially Th1 responses, and thereby limit infection-induced immunopathology (Mills, 2004). Alternatively their induction may serve as an evasion strategy by the bacteria to suppress protective Th1 responses.

Another population of suppressor T cells, initially characterized through their expression of CD25, but more recently by the transcriptional repressor Foxp3, termed natural Treg cells also play a crucial role in regulating immune responses (Hori and Sakaguchi, 2004). These cells which account for 5–50 per cent of circulating T cells were shown by depletion and adoptive transfer experiments to be capable of preventing autoimmune diseases through their ability to maintain tolerance to self antigens (Sakaguchi, 2000). $CD25^+$ T cells have been shown in a number of systems to be capable of suppressing proliferation and cytokine secretion by $CD4^+CD25^-$ T cells. Natural Treg cells also play a role in mucosal tolerance and help to control immune responses to inhaled antigens in the lung (Ostroukhova *et al.*, 2004) and to commensal bacteria in the gut (Powrie, 2004). In addition there is evidence that they function to regulate protective immunity to pathogens, including pathogenic bacteria (Mills, 2004).

10.5.2 Treg cells induced by bacteria modulated DC

It has been suggested that the activation status of the DC, rather than the lineage, determines its ability to selectively promote T cell subtypes (Barrat *et al.*, 2002). Many TLR ligands, including LPS, CpG motifs in bacterial DNA and viral dsRNA have been shown to promote IL-12 production, and to activate DC that promote the differentiation of Th1 cells (Hemmi *et al.*, 2000; Moser and Murphy, 2000) (Figure 10.1). In contrast, products of helminth parasites, yeast hyphae and CT activate DCs, which directs the induction of Th2 cells (d'Ostiani *et al.*, 2000; Gagliardi *et al.*, 2000; Moser and Murphy, 2000). Since inducible Treg cells (Tr1 and Th3 cells) arise from naive or resting $CD4^+$ T cells in the periphery, it was highly conceivable that DCs activated with an appropriate stimulus, such as certain pathogen-derived molecules, could selectively promote the induction of Treg cells. Indeed, evidence is emerging that bacteria or bacteria-derived molecules can promote the induction of Tr1 cells, through their interaction with DCs or other innate cells, such as macrophages.

The induction of high levels of IL-10 and/or TGF-β by DCs following infection has been reported for a number of bacterial species. In certain cases, specific pathogen-derived immunoregulatory molecules have been identified that stimulate IL-10-producing DCs. *B. pertussis*, the etiologic agent of whooping cough, causes a severe and protracted respiratory disease, often complicated by secondary infections that can have a lethal outcome in young children. Recovery from infection in humans and

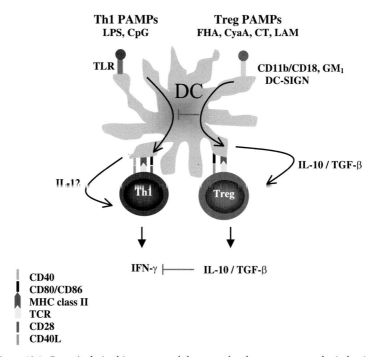

Th1 PAMPs
LPS, CpG

Treg PAMPs
FHA, CyaA, CT, LAM

TLR

CD11b/CD18, GM$_1$
DC-SIGN

DC

IL-10 / TGF-β

IL-12

Th1

Treg

IFN-γ ⊢——— IL-10 / TGF-β

CD40
CD80/CD86
MHC class II
TCR
CD28
CD40L

Figure 10.1. Bacteria-derived immunomodulatory molecules can promote the induction of Th1 or Treg cells by modulating DC activation. Bacteria-derived immunomodulatory molecules, including LPS and CpG motifs in bacterial DNA, that bind to Toll-like receptors (TLR) on dendritic cells (DCs) activate signaling pathways that enhance MHC class II, CD80, CD86 and CD40 expression and IL-12 production by DCs, promote the induction of Th1 cells from naive T cells. In contrast a distinct set of bacteria-derived immunomodulatory molecules, including FHA and CyaA from *B. pertussis*, LAM from *M. tuberculosis* and cholera toxin (CT) that bind to other PPR, including CD11b/CD18, GM$_1$ or DC-SIGN, activate signaling pathways that enhance MHC class II and CD80 expression and IL-10 production, but suppress CD40 expression and IL-12 production by DCs, promote the induction of Treg cells from naive T cells. These Treg cells suppress immune responses mediated by Th1 cells.

mice is associated with the development of *B. pertussis*-specific Th1 cells (Mills *et al.*, 1993; Ryan *et al.*, 1997). However, antigen-specific Th1 immune responses in the lungs of infected mice are severely suppressed during the acute phase of infection (McGuirk *et al.*, 1998). The *B. pertussis* virulence factors, FHA and CyaA, have been shown to inhibit IL-12 and enhance IL-10 production from macrophages and DCs and as a result selectively stimulate the induction of IL-10 secreting Tr1 cells from naive

T cells (McGuirk *et al.*, 2002; McGuirk and Mills, 2000; Boyd *et al.*, 2005; Ross *et al.*, 2004). Furthermore, Tr1 clones specific for FHA have been generated from the lungs of acutely infected mice, but could not be generated from the spleens of infected or convalescent mice, or the lungs of naive mice (McGuirk *et al.*, 2002).

In a murine model of eosinophilic airway inflammation, treatment with *M. vaccae* has been demonstrated to have suppressive activity on airway eosinophilia (Zuany-Amorim *et al.*, 2002). Treatment of mice with killed *M. vaccae* suspension gives rise to $CD11c^+$ cells that express high levels of IL-10, TGF-β and IFN-α mRNA (Adams *et al.*, 2004), cytokines which are thought to strongly promote the differentiation of Treg cells. Indeed, *M. vaccae* treatment has been shown to promote the induction of allergen-specific $CD4^+CD45RB^{low}$ Treg cells, which not only confer protection against airway epithelium damage but also reduce other pathophysiological readouts such as bronchial hyperresponsiveness (Zuany-Amorim *et al.*, 2002).

10.6 ROLE OF INDUCIBLE AND NATURAL TREG CELLS IN BACTERIAL INFECTIONS

There is now mounting evidence that numerous bacteria and their products selectively promote the differentiation of Treg as a deliberate strategy to misdirect the immune response and thus facilitate their persistence in vivo. As previously discussed, the respiratory pathogen *B. pertussis* causes a protracted infection associated with suppressed Th1 responses during the acute phase of infection. Tr1 clones specific for FHA and pertactin have been generated from the lungs of acutely infected mice. (McGuirk *et al.*, 2002).

Treg cells may also play a significant role in the persistence of chronic bacterial infections. In tuberculosis, 15 per cent of infected patients fail to respond to intradermal injection with purified protein derivative (PPD) (Boussiotis *et al.*, 2000). T cells from patients with normal DTH responses proliferate and secrete IFN-γ. In contrast, T cells from unresponsive patients proliferate poorly and secrete high levels of IL-10 (Boussiotis *et al.*, 2000; Delgado *et al.*, 2002). Furthermore, recent studies in IL-10 transgenic mice demonstrate that suppression of protective Th1 responses and increased susceptibility to reactivation of tuberculosis is strongly influenced by the expression of IL-10 during the latent phase of infection (Turner *et al.*, 2002). In addition, some bacterial products have been shown to be capable of directly inducing Treg cells. *Streptococcus pyogenes*, a member of the group A streptococci, causes numerous diseases in humans, including pharyngitis, skin infections and post infectious rheumatic fever. One of the major virulence factors of *S. pyogenes* is the M protein, which is thought to play a role in bacterial adhesion. However, a recent study has demonstrated that M protein interacts with CD46 on T cells and directly promotes the differentiation of IL-10 secreting Treg cells from naive precursors (Price *et al.*, 2005).

Although the induction of Treg cells following infection in many cases promotes bacterial persistence, bacterial-specific Treg cells are not exclusively detrimental for the host. The intestine contains a large quantity of resident bacteria that are a significant source of both antigen and pro-inflammatory molecules. However, despite these immune responses the intestine remains in a state of controlled inflammation, suggesting that mucosal immune responses to enteric bacteria are tightly regulated. Support for this concept has come from a number of experimental systems demonstrating that altered regulation of intestinal T cell function can result in chronic inflammatory gut disorders. In the murine model of inflammatory bowel disease, adoptive transfer of $CD4^+CD45RB^{high}$ cells from normal mice to SCID recipients results in severe colitis, wasting and death. However, co-transfer of the reciprocal $CD4^+CD45RB^{low}$ subset prevents disease, due to the presence of Treg cells within this subset that secrete high levels of IL-10 and TGF-β (Powrie *et al.*, 1994). Furthermore, it is now well established that IL-10 deficient mice not housed under germ free or specific pathogen-free conditions develop spontaneous colitis (Powrie *et al.*, 1996). Therefore, one potentially important immunoregulatory function of certain bacterial species is their involvement in the generation of distinct T cell subtypes and maintenance of immune homeostasis.

Direct evidence for the role of bacteria-specific Treg cells in the maintenance of homeostasis has come from recent studies on a common gut pathogen *H. hepaticus* encountered in animal facilities. *H. hepaticus* infection induced a population of $CD4^+CD45RB^{low}$ Treg cells that inhibited the development of colitis triggered by this Gram-negative bacterium in IL-10 knockout mice (Kullberg *et al.*, 2001, 2002). The protective Treg cells from *H. hepaticus* infected mice were enriched within the $CD25^-$ $CD45RB^{low}$ subset, but absent from naive mice, suggesting that these Treg cells, rather than being endogenous, represent a memory population resulting from previous exposure to bacterial antigen. The major question that arises from this study is whether gut flora-specific Treg cells, similar to those described above, prevent the appearance of inflammatory bowel disease in humans. Evidence in support of this possibility comes from studies demonstrating T cell tolerance in the intestine toward resident gut flora is mediated by bacterial antigen-specific CD4$^+$ T cells that secrete IL-10 and TGF-β (Khoo *et al.*, 1997). In addition, it has also been suggested that suppressed CD4$^+$ memory T cell responsiveness from *H. pylori* infected individuals is due to the presence of pathogen-specific Treg cells (Lundgren *et al.*, 2003). It has also been suggested that the primary role of natural CD4$^+$CD25$^+$ Treg cells is to limit immune mediated pathology.

Another example of a beneficial role for regulatory cells in controlling exaggerated inflammatory immune responses has come from studies on infection of Toll-like receptor (TLR)-4 deficient mice with the respiratory pathogen *B. pertussis*. During the acute phase of infection, Th1 responses were shown to be enhanced and IL-10 producing T cells significantly reduced in TLR-4 deficient mice. This was associated with enhanced inflammatory cytokine production, cellular infiltration and severe pathological changes in the lungs of the TLR-4 deficient mice (Higgins *et al.*, 2003). These findings suggest that ligation of conserved pathogen molecules to PRRs promotes the induction of Treg cells that prevents immune-mediated pathology. Therefore, IL-10 production from innate cells leading to the differentiation of Treg cells, may represent a protective strategy adopted by the host to limit collateral damage mediated by excessive pathogen-stimulated inflammatory responses.

10.7 CONCLUSIONS

The immune system is capable of mounting an array of innate and adaptive responses that can prevent or control bacterial diseases and confer immunity to re-infection with pathogenic bacteria. However, protective immune responses, especially those against intracellular pathogens can cause collateral damage to host tissues and must be tightly regulated to prevent excessive immunopathology. Furthermore, pathogens have evolved a variety of immune subversion approaches for subverting protective immunity and prolonging their survival in the host. Bacteria and their products can induce regulatory cytokines, especially IL-10 and TGF-β, from innate immune cells, including macrophages and DCs, which in turn can generate bacteria-specific Treg cells. There is growing evidence of inducible Treg cells in bacterial infections, which either suppress effector immune response and thereby delay or prevent pathogen clearance, or allow pathogen elimination with limited inflammatory pathology. Treg cells and anti-inflammatory cytokines play a central role in regulating immune responses to self and foreign antigens, and a better understanding of their role in immunity during bacterial infection should help in the design of new or improved vaccines and therapies against many diseases, that are still responsible for high levels of morbidity and mortality worldwide.

REFERENCES

Adams, V. C., Hunt, J. R., Martinelli, R., Palmer, R., Rook, G. A., and Brunet, L. R. (2004). *Mycobacterium vaccae* induces a population of pulmonary CD11c$^+$ cells with regulatory potential in allergic mice. *Eur. J. Immunol.* **34**, 631−8.

Agrawal, A., Lingappa, J., Leppla, S. H., Agrawal, S., Jabbar, A., Quinn, C., and Pulendran, B. (2003). Impairment of dendritic cells and adaptive immunity by anthrax lethal toxin. *Nature* **424**, 329−34.

al-Ramadi, B. K., Brodkin, M. A., Mosser, D. M., and Eisenstein, T. K. (1991a). Immunosuppression induced by attenuated *Salmonella*. Evidence for mediation by macrophage precursors. *J. Immunol.* **146**, 2737−46.

al-Ramadi, B. K., Chen, Y. W., Meissler, J. J., Jr., and Eisenstein, T. K. (1991b). Immunosuppression induced by attenuated *Salmonella*. Reversal by IL-4. *J. Immunol.* **147**, 1954−61.

Anjuere, F., Luci, C., Lebens, M., Rousseau, D., Hervouet, C., Milon, G., Holmgren, J., Ardavin, C., and Czerkinsky, C. (2004). In vivo adjuvant-induced mobilization and maturation of gut dendritic cells after oral administration of cholera toxin. *J. Immunol.* **173**, 5103−11.

Bagley, K. C., Abdelwahab, S. F., Tuskan, R. G., Fouts, T. R., and Lewis, G. K. (2002a). Cholera toxin and heat-labile enterotoxin activate human monocyte-derived dendritic cells and dominantly inhibit cytokine production through a cyclic AMP-dependent pathway. *Infect. Immun.* **70**, 5533−9.

Bagley, K. C., Abdelwahab, S. F., Tuskan, R. G., Fouts, T. R., and Lewis, G. K. (2002b). Pertussis toxin and the adenylate cyclase toxin from *Bordetella pertussis* activate human monocyte-derived dendritic cells and dominantly inhibit cytokine production through a cAMP-dependent pathway. *J. Leukoc. Biol.* **72**, 962−9.

Bagley, K. C., Abdelwahab, S. F., Tuskan, R. G., and Lewis, G. K. (2005). *Pasteurella multocida* toxin activates human monocyte-derived and murine bone marrow-derived dendritic cells in vitro but suppresses antibody production in vivo. *Infect. Immun.* **73**, 413−21.

Balkhi, M. Y., Sinha, A., and Natarajan, K. (2004). Dominance of CD86, transforming growth factor- beta 1, and interleukin-10 in *Mycobacterium tuberculosis* secretory antigen-activated dendritic cells regulates T helper 1 responses to mycobacterial antigens. *J. Infect. Dis.* **189**, 1598–609.

Bancereau, J., Briere, F., Caux, C., Davoust, J., Lebecque, S., Liu, Y. J., Pulendran, B., and Palucka, K. (2000). Immunobiology of dendritic cells. *Annu. Rev. Immunol.* **18**, 767–811.

Barnes, P. F., Chatterjee, D., Abrams, J. S., Lu, S., Wang, E., Yamamura, M., Brennan, P. J., and Modlin, R. L. (1992). Cytokine production induced by *Mycobacterium tuberculosis* lipoarabinomannan. Relationship to chemical structure. *J. Immunol.* **149**, 541–7.

Barrat, F. J., Cua, D. J., Boonstra, A., Richards, D. F., Crain, C., Savelkoul, H. F., de Waal-Malefyt, R., Coffman, R. L., Hawrylowicz, C. M., and O'Garra, A. (2002). In vitro generation of interleukin 10-producing regulatory CD4$^{(+)}$ T cells is induced by immunosuppressive drugs and inhibited by T helper type 1 (Th1)- and Th2-inducing cytokines. *J. Exp. Med.* **195**, 603–16.

Beuscher, H. U., Rodel, F., Forsberg, A., and Rollinghoff, M. (1995). Bacterial evasion of host immune defense: *Yersinia enterocolitica* encodes a suppressor for tumor necrosis factor alpha expression. *Infect. Immun.* **63**, 1270–7.

Boirivant, M., Fuss, I. J., Ferroni, L., De Pascale, M., and Strober, W. (2001). Oral administration of recombinant cholera toxin subunit B inhibits IL-12-mediated murine experimental (trinitrobenzene sulfonic acid) colitis. *J. Immunol.* **166**, 3522–32.

Bonecini-Almeida, M. G., Ho, J. L., Boechat, N., Huard, R. C., Chitale, S., Doo, H., Geng, J., Rego, L., Lazzarini, L. C., Kritski, A. L., Johnson, W. D., Jr., McCaffrey, T. A., and Silva, J. R. (2004). Down-modulation of lung immune responses by interleukin-10 and transforming growth factor beta (TGF-beta) and analysis of TGF-beta receptors I and II in active tuberculosis. *Infect. Immun.* **72**, 2628–34.

Boussiotis, V. A., Tsai, E. Y., Yunis, E. J., Thim, S., Delgado, J. C., Dascher, C. C., Berezovskaya, A., Rousset, D., Reynes, J. M., and Goldfeld, A. E. (2000). IL-10-producing T cells suppress immune responses in anergic tuberculosis patients. *J. Clin. Invest.* **105**, 1317–25.

Boyd, A. P., Ross, P. J., Conroy, H., Mahon, N., Lavelle, E. C., and Mills, K. H. (2005). *Bordetella pertussis* adenylate cyclase toxin modulates innate and adaptive immune responses: distinct roles for acylation and enzymatic activity in immunomodulation and cell death. *J. Immunol.* **175**, 730–8.

Braun, M. C., He, J., Wu, C. Y., and Kelsall, B. L. (1999). Cholera toxin suppresses interleukin (IL)-12 production and IL-12 receptor beta1 and beta2 chain expression. *J. Exp. Med.* **189**, 541–52.

Brzoza, K. L., Rockel, A. B., and Hiltbold, E. M. (2004). Cytoplasmic entry of *Listeria monocytogenes* enhances dendritic cell maturation and T cell differentiation and function. *J. Immunol.* **173**, 2641–51.

Byrne, P., McGuirk, P., Todryk, S., and Mills, K. H. (2004). Depletion of NK cells results in disseminating lethal infection with *Bordetella pertussis* associated with a reduction of antigen-specific Th1 and enhancement of Th2, but not Tr1 cells. *Eur. J. Immunol.* **34**, 2579–88.

Chen, W., Shu, D., and Chadwick, V. S. (2001). *Helicobacter pylori* infection: mechanism of colonization and functional dyspepsia reduced colonization of gastric mucosa by *Helicobacter pylori* in mice deficient in interleukin-10. *J. Gastroenterol. Hepatol.* **16**, 377–83.

Chiao, J. W., Villalon, P., Schwartz, I., and Wormser, G. P. (2000). Modulation of lymphocyte proliferative responses by a canine Lyme disease vaccine of recombinant outer surface protein A (OspA). *FEMS Immunol. Med. Microbiol.* **28**, 193–6.

Chieppa, M., Bianchi, G., Doni, A., Del Prete, A., Sironi, M., Laskarin, G., Monti, P., Piemonti, L., Biondi, A., Mantovani, A., Introna, M., and Allavena, P. (2003). Cross-linking of the mannose receptor on monocyte-derived dendritic cells activates an anti-inflammatory immunosuppressive program. *J. Immunol.* **171**, 4552–60.

Christensen, H. R., Frokiaer, H., and Pestka, J. J. (2002). Lactobacilli differentially modulate expression of cytokines and maturation surface markers in murine dendritic cells. *J. Immunol.* **168**, 171–8.

Dahl, K. E., Shiratsuchi, H., Hamilton, B. D., Ellner, J. J., and Toossi, Z. (1996). Selective induction of transforming growth factor beta in human monocytes by lipoarabinomannan of *Mycobacterium tuberculosis*. *Infect. Immun.* **64**, 399–405.

de Souza, M. S., Smith, A. L., Beck, D. S., Terwilliger, G. A., Fikrig, E., and Barthold, S. W. (1993). Long-term study of cell-mediated responses to *Borrelia burgdorferi* in the laboratory mouse. *Infect. Immun.* **61**, 1814–22.

Deckert, M., Soltek, S., Geginat, G., Lutjen, S., Montesinos-Rongen, M., Hof, H., and Schluter, D. (2001). Endogenous interleukin-10 is required for prevention of a hyperinflammatory intracerebral immune response in *Listeria monocytogenes* meningoencephalitis. *Infect. Immun.* **69**, 4561–71.

Delgado, J. C., Tsai, E. Y., Thim, S., Baena, A., Boussiotis, V. A., Reynes, J. M., Sath, S., Grosjean, P., Yunis, E. J., and Goldfeld, A. E. (2002). Antigen-specific and persistent tuberculin anergy in a cohort of pulmonary tuberculosis patients from rural Cambodia. *Proc. Natl Acad. Sci. U S A* **99**, 7576–81.

Demangel, C., Bertolino, P., and Britton, W. J. (2002). Autocrine IL-10 impairs dendritic cell (DC)-derived immune responses to mycobacterial infection by suppressing DC trafficking to draining lymph nodes and local IL-12 production. *Eur. J. Immunol.* **32**, 994–1002.

d'Ostiani, C. F., Del Sero, G., Bacci, A., Montagnoli, C., Spreca, A., Mencacci, A., Ricciardi-Castagnoli, P., and Romani, L. (2000). Dendritic cells discriminate between yeasts and hyphae of the fungus *Candida albicans*. Implications for initiation of T helper cell immunity in vitro and in vivo. *J. Exp. Med.* **191**, 1661–74.

Erfurth, S. E., Grobner, S., Kramer, U., Gunst, D. S., Soldanova, I., Schaller, M., Autenrieth, I. B., and Borgmann, S. (2004). *Yersinia enterocolitica* induces apoptosis and inhibits surface molecule expression and cytokine production in murine dendritic cells. *Infect. Immun.* **72**, 7045–54.

Gagliardi, M. C., Sallusto, F., Marinaro, M., Langenkamp, A., Lanzavecchia, A., and De Magistris, M. T. (2000). Cholera toxin induces maturation of human dendritic cells and licences them for Th2 priming. *Eur. J. Immunol.* **30**, 2394–403.

Gagliardi, M. C., Teloni, R., Giannoni, F., Pardini, M., Sargentini, V., Brunori, L., Fattorini, L., and Nisini, R. (2005). *Mycobacterium bovis* Bacillus Calmette-Guerin infects DC-SIGN-dendritic cell and causes the inhibition of IL-12 and the enhancement of IL-10 production. *J. Leukoc. Biol.* **78**, 106–13.

Geijtenbeek, T. B., Van Vliet, S. J., Koppel, E. A., Sanchez-Hernandez, M., Vandenbroucke-Grauls, C. M., Appelmelk, B., and Van Kooyk, Y. (2003). Mycobacteria target DC-SIGN to suppress dendritic cell function. *J. Exp. Med.* **197**, 7–17.

Happel, K. I., Zheng, M., Young, E., Quinton, L. J., Lockhart, E., Ramsay, A. J., Shellito, J. E., Schurr, J. R., Bagby, G. J., Nelson, S., and Kolls, J. K. (2003). Cutting edge: roles of Toll-like receptor 4 and IL-23 in IL-17 expression in response to *Klebsiella pneumoniae* infection. *J. Immunol.* **170**, 4432–6.

Hemmi, H., Takeuchi, O., Kawai, T., Kaisho, T., Sato, S., Sanjo, H., Matsumoto, M., Hoshino, K., Wagner, H., Takeda, K., and Akira, S. (2000). A Toll-like receptor recognizes bacterial DNA. *Nature* **408**, 740–5.

Higgins, S. C., Lavelle, E. C., McCann, C., Keogh, B., McNeela, E., Byrne, P., O'Gorman, B., Jarnicki, A., McGuirk, P., and Mills, K. H. (2003). Toll-like receptor 4-mediated innate IL-10 activates antigen-specific regulatory T cells and confers resistance to *Bordetella pertussis* by inhibiting inflammatory pathology. *J. Immunol.* **171**, 3119–27.

Hoover, D. L., Friedlander, A. M., Rogers, L. C., Yoon, I. K., Warren, R. L., and Cross, A. S. (1994). Anthrax edema toxin differentially regulates lipopolysaccharide-induced monocyte production of tumor necrosis factor alpha and interleukin-6 by increasing intracellular cyclic AMP. *Infect. Immun.* **62**, 4432–9.

Hori, S. and Sakaguchi, S. (2004). Foxp3: a critical regulator of the development and function of regulatory T cells. *Microbes Infect.* **6**, 745–51.

Ishibashi, Y., Claus, S., and Relman, D. A. (1994). *Bordetella pertussis* filamentous hemagglutinin interacts with a leukocyte signal transduction complex and stimulates bacterial adherence to monocyte CR3 (CD11b/CD18). *J. Exp. Med.* **180**, 1225–33.

Jacobs, M., Brown, N., Allie, N., Gulert, R., and Ryffel, B. (2000). Increased resistance to mycobacterial infection in the absence of interleukin-10. *Immunology* **100**, 494–501.

Jacobs, M., Fick, L., Allie, N., Brown, N., and Ryffel, B. (2002). Enhanced immune response in *Mycobacterium bovis* bacille Calmette Guerin (BCG)-infected IL-10-deficient mice. *Clin. Chem. Lab. Med.* **40**, 893–902.

Janeway, C. A., Jr. and Medzhitov, R. (2002). Innate immune recognition. *Annu. Rev. Immunol.* **20**, 197–216.

Kalinski, P., Hilkens, C. M., Wierenga, E. A., and Kapsenberg, M. L. (1999). T-cell priming by type-1 and type-2 polarized dendritic cells: the concept of a third signal. *Immunol. Today* **20**, 561–7.

Kaplan, G., Gandhi, R. R., Weinstein, D. E., Levis, W. R., Patarroyo, M. E., Brennan, P. J., and Cohn, Z. A. (1987). *Mycobacterium leprae* antigen-induced suppression of T cell proliferation in vitro. *J. Immunol.* **138**, 3028–34.

Khoo, U. Y., Proctor, I. E., and Macpherson, A. J. (1997). CD4$^+$ T cell down-regulation in human intestinal mucosa: evidence for intestinal tolerance to luminal bacterial antigens. *J. Immunol.* **158**, 3626–34.

Knipp, U., Birkholz, S., Kaup, W., Mahnke, K., and Opferkuch, W. (1994). Suppression of human mononuclear cell response by *Helicobacter pylori*: effects on isolated monocytes and lymphocytes. *FEMS Immunol. Med. Microbiol.* **8**, 157–66.

Kullberg, M. C., Jankovic, D., Gorelick, P. L., Caspar, P., Letterio, J. J., Cheever, A. W., and Sher, A. (2002). Bacteria-triggered CD4$^+$ T regulatory cells suppress *Helicobacter hepaticus*-induced colitis. *J. Exp. Med.* **196**, 505–15.

Kullberg, M. C., Rothfuchs, A. G., Jankovic, D., Caspar, P., Wynn, T. A., Gorelick, P. L., Cheever, A. W., and Sher, A. (2001). *Helicobacter hepaticus*-induced colitis in interleukin-10-deficient mice: cytokine requirements for the induction and maintenance of intestinal inflammation. *Infect. Immun.* **69**, 4232–41.

Lanzavecchia, A. and Sallusto, F. (2001). Regulation of T cell immunity by dendritic cells. *Cell* **106**, 263–6.

Lavelle, E. C., McNeela, E., Armstrong, M. E., Leavy, O., Higgins, S. C., and Mills, K. H. (2003). Cholera toxin promotes the induction of regulatory T cells specific for bystander antigens by modulating dendritic cell activation. *J. Immunol.* **171**, 2384–92.

Lopes, L. M., Maroof, A., Dougan, G., and Chain, B. M. (2000). Inhibition of T-cell response by *Escherichia coli* heat-labile enterotoxin-treated epithelial cells. *Infect. Immun.* **68**, 6891–5.

Lundgren, A., Suri-Payer, E., Enarsson, K., Svennerholm, A. M., and Lundin, B. S. (2003). *Helicobacter pylori*-specific CD4$^+$ CD25^{high} regulatory T cells suppress memory T-cell responses to *H. pylori* in infected individuals. *Infect. Immun.* **71**, 1755–62.

MacFarlane, A. S., Schwacha, M. G., and Eisenstein, T. K. (1999). In vivo blockage of nitric oxide with aminoguanidine inhibits immunosuppression induced by an attenuated strain of *Salmonella typhimurium*, potentiates *Salmonella* infection, and inhibits macrophage and polymorphonuclear leukocyte influx into the spleen. *Infect. Immun.* **67**, 891–8.

Mahon, B. P., Sheahan, B. J., Griffin, F., Murphy, G., and Mills, K. H. (1997). Atypical disease after *Bordetella pertussis* respiratory infection of mice with targeted disruptions of interferon-gamma receptor or immunoglobulin mu chain genes. *J. Exp. Med.* **186**, 1843–51.

McBride, J. M., Jung, T., de Vries, J. E., and Aversa, G. (2002). IL-10 alters DC function via modulation of cell surface molecules resulting in impaired T-cell responses. *Cell Immunol.* **215**, 162–72.

McGuirk, P., Johnson, P. A., Ryan, E. J., and Mills, K. H. (2000). Filamentous hemagglutinin and pertussis toxin from *Bordetella pertussis* modulate immune responses to unrelated antigens. *J. Infect. Dis.* **182**, 1286–9.

McGuirk, P., Mahon, B. P., Griffin, F., and Mills, K. H. (1998). Compartmentalization of T cell responses following respiratory infection with *Bordetella pertussis*: hyporesponsiveness of lung T cells is associated with modulated expression of the co-stimulatory molecule CD28. *Eur. J. Immunol.* **28**, 153–63.

McGuirk, P., McCann, C., and Mills, K. H. (2002). Pathogen-specific T
    regulatory 1 cells induced in the respiratory tract by a bacterial molecule
    that stimulates interleukin 10 production by dendritic cells: a novel
    strategy for evasion of protective T helper type 1 responses by *Bordetella
    pertussis*. *J. Exp. Med.* **195**, 221–31.

McGuirk, P. and Mills, K. H. (2000). Direct anti-inflammatory effect of a bacterial
    virulence factor: IL-10-dependent suppression of IL-12 production by
    filamentous hemagglutinin from *Bordetella pertussis*. *Eur. J. Immunol.*
    30, 415–22.

Mehra, V., Brennan, P. J., Rada, E., Convit, J., and Bloom, B. R. (1984).
    Lymphocyte suppression in leprosy induced by unique *M. leprae*
    glycolipid. *Nature* **308**, 194–6.

Mills, K. H. (2001). Immunity to *Bordetella pertussis*. *Microbes Infect.* **3**,
    655–77.

Mills, K. H. (2004). Regulatory T cells: friend or foe in immunity to
    infection? *Nat. Rev. Immunol.* **4**, 841–55.

Mills, K. H., Barnard, A., Watkins, J., and Redhead, K. (1993). Cell-mediated
    immunity to *Bordetella pertussis*: role of Th1 cells in bacterial clearance
    in a murine respiratory infection model. *Infect. Immun.* **61**, 399–410.

Modlin, R. L. (1994). Th1–Th2 paradigm: insights from leprosy. *J. Invest.
    Dermatol.* **102**, 828–32.

Moore, K. W., de Waal Malefyt, R., Coffman, R. L., and O'Garra, A. (2001).
    Interleukin-10 and the interleukin-10 receptor. *Annu. Rev. Immunol.*
    19, 683–765.

Moser, M. and Murphy, K. M. (2000). Dendritic cell regulation of TH1–TH2
    development. *Nat. Immunol.* **1**, 199–205.

Moura, A. C., Modolell, M., and Mariano, M. (1997). Down-regulatory effect
    of *Mycobacterium leprae* cell wall lipids on phagocytosis, oxidative
    respiratory burst and tumour cell killing by mouse bone marrow
    derived macrophages. *Scand. J. Immunol.* **46**, 500–5.

Nakane, A., Asano, M., Sasaki, S., Nishikawa, S., Miura, T., Kohanawa, M.,
    and Minagawa, T. (1996). Transforming growth factor beta is protective
    in host resistance against *Listeria monocytogenes* infection in mice.
    Infect. Immun. **64**, 3901–4.

Natarajan, K., Latchumanan, V. K., Singh, B., Singh, S., and Sharma, P. (2003).
    Down-regulation of T helper 1 responses to mycobacterial antigens due
    to maturation of dendritic cells by 10-kDa *Mycobacterium tuberculosis*
    secretory antigen. *J. Infect. Dis.* **187**, 914–28.

Nigou, J., Zelle-Rieser, C., Gilleron, M., Thurnher, M., and Puzo, G. (2001).
    Mannosylated lipoarabinomannans inhibit IL-12 production by human

dendritic cells: evidence for a negative signal delivered through the mannose receptor. *J. Immunol.* **166**, 7477–85.

Ostroukhova, M., Seguin-Devaux, C., Oriss, T. B., Dixon-McCarthy, B., Yang, L., Ameredes, B. T., Corcoran, T. E., and Ray, A. (2004). Tolerance induced by inhaled antigen involves CD4$^{(+)}$ T cells expressing membrane-bound TGF-beta and FOXP3. *J. Clin. Invest.* **114**, 28–38.

Othieno, C., Hirsch, C. S., Hamilton, B. D., Wilkinson, K., Ellner, J. J., and Toossi, Z. (1999). Interaction of *Mycobacterium tuberculosis*-induced transforming growth factor beta1 and interleukin-10. *Infect. Immun.* **67**, 5730–5.

Pena, J. A., Li, S. Y., Wilson, P. H., Thibodeau, S. A., Szary, A. J., and Versalovic, J. (2004). Genotypic and phenotypic studies of murine intestinal lactobacilli: species differences in mice with and without colitis. *Appl. Environ. Microbiol.* **70**, 558–68.

Pena, J. A., Rogers, A. B., Ge, Z., Ng, V., Li, S. Y., Fox, J. G., and Versalovic, J. (2005). Probiotic *Lactobacillus* spp. diminish *Helicobacter hepaticus*-induced inflammatory bowel disease in interleukin-10-deficient mice. *Infect. Immun.* **73**, 912–20.

Powrie, F. (2004). Immune regulation in the intestine: a balancing act between effector and regulatory T cell responses. *Ann. N Y Acad. Sci.* **1029**, 132–41.

Powrie, F., Carlino, J., Leach, M. W., Mauze, S., and Coffman, R. L. (1996). A critical role for transforming growth factor-β but not interleukin 4 in the suppression of T helper type 1-mediated colitis by CD45RB^{low} CD4$^+$ T cells. *J. Exp. Med.* **183**, 2669–74.

Powrie, F., Correa-Oliveira, R., Mauze, S., and Coffman, R. L. (1994). Regulatory interactions between CD45RBhigh and CD45RBlow CD4$^+$ T cells are important for the balance between protective and pathogenic cell-mediated immunity. *J. Exp. Med.* **179**, 589–600.

Price, J. D., Schaumburg, J., Sandin, C., Atkinson, J. P., Lindahl, G., and Kemper, C. (2005). Induction of a regulatory phenotype in human CD4$^+$ T cells by streptococcal M protein. *J. Immunol.* **175**, 677–84.

Redpath, S., Ghazal, P., and Gascoigne, N. R. (2001). Hijacking and exploitation of IL-10 by intracellular pathogens. *Trends. Microbiol.* **9**, 86–92.

Reis e Sousa, C. (2004). Activation of dendritic cells: translating innate into adaptive immunity. *Curr. Opin. Immunol.* **16**, 21–5.

Rosenfeldt, V., Benfeldt, E., Nielsen, S. D., Michaelsen, K. F., Jeppesen, D. L., Valerius, N. H., and Paerregaard, A. (2003). Effect of probiotic *Lactobacillus* strains in children with atopic dermatitis. *J. Allergy Clin. Immunol.* **111**, 389–95.

Ross, P. J., Lavelle, E. C., Mills, K. H., and Boyd, A. P. (2004). Adenylate cyclase toxin from *Bordetella pertussis* synergizes with lipopolysaccharide to promote innate interleukin-10 production and enhances the induction of Th2 and regulatory T cells. *Infect. Immun.* **72**, 1568–79.

Ryan, E. J., McNeela, E., Pizza, M., Rappuoli, R., O'Neill, L., and Mills, K. H. (2000). Modulation of innate and acquired immune responses by *Escherichia coli* heat-labile toxin: distinct pro- and anti-inflammatory effects of the nontoxic AB complex and the enzyme activity. *J. Immunol.* **165**, 5750–9.

Ryan, M., Murphy, G., Gothefors, L., Nilsson, L., Storsaeter, J., and Mills, K. H. (1997). *Bordetella pertussis* respiratory infection in children is associated with preferential activation of type 1 T helper cells. *J. Infect. Dis.* **175**, 1246–50.

Sakaguchi, S. (2000). Regulatory T cells: key controllers of immunologic self-tolerance. *Cell* **101**, 455–8.

Salgame, P. R., Birdi, T. J., Lad, S. J., Mahadevan, P. R., and Antia, N. H. (1984). Mechanism of immunosuppression in leprosy – macrophage membrane alterations. *J. Clin. Lab. Immunol.* **14**, 145–9.

Schaible, U. E., and Kaufmann, S. H. (2000). CD1 and CD1-restricted T cells in infections with intracellular bacteria. *Trends Microbiol.* **8**, 419–25.

Schoppet, M., Bubert, A., and Huppertz, H. I. (2000). Dendritic cell function is perturbed by *Yersinia enterocolitica* infection in vitro. *Clin. Exp. Immunol.* **122**, 316–23.

Sewnath, M. E., Olszyna, D. P., Birjmohun, R., ten Kate, F. J., Gouma, D. J., and van Der Poll, T. (2001). IL-10-deficient mice demonstrate multiple organ failure and increased mortality during *Escherichia coli* peritonitis despite an accelerated bacterial clearance. *J. Immunol.* **166**, 6323–31.

Shannon, J. G., Howe, D., and Heinzen, R. A. (2005). Virulent *Coxiella burnetii* does not activate human dendritic cells: role of lipopolysaccharide as a shielding molecule. *Proc. Natl Acad. Sci. U S A* **102**, 8722–7.

Sing, A., Roggenkamp, A., Geiger, A. M., and Heesemann, J. (2002). *Yersinia enterocolitica* evasion of the host innate immune response by V antigen-induced IL-10 production of macrophages is abrogated in IL-10-deficient mice. *J. Immunol.* **168**, 1315–21.

Smits, H. H., Engering, A., van der Kleij, D., de Jong, E. C., Schipper, K., van Capel, T. M., Zaat, B. A., Yazdanbakhsh, M., Wierenga, E. A., van Kooyk, Y., and Kapsenberg, M. L. (2005). Selective probiotic bacteria induce IL-10-producing regulatory T cells in vitro by modulating dendritic cell function through dendritic cell-specific intercellular adhesion molecule 3-grabbing nonintegrin. *J. Allergy Clin. Immunol.* **115**, 1260–7.

Tournier, J. N., Quesnel-Hellmann, A., Mathieu, J., Montecucco, C., Tang, W. J., Mock, M., Vidal, D. R., and Goossens, P. L. (2005). Anthrax edema toxin cooperates with lethal toxin to impair cytokine secretion during infection of dendritic cells. *J. Immunol.* **174**, 4934–41.

Turner, J., Gonzalez-Juarrero, M., Ellis, D. L., Basaraba, R. J., Kipnis, A., Orme, I. M., and Cooper, A. M. (2002). In vivo IL-10 production reactivates chronic pulmonary tuberculosis in C57BL/6 mice. *J. Immunol.* **169**, 6343–51.

van Kooyk, Y. and Geijtenbeek, T. B. (2003). DC-SIGN: escape mechanism for pathogens. *Nat. Rev. Immunol.* **3**, 697–709.

Yao, T., Mecsas, J., Healy, J. I., Falkow, S., and Chien, Y. (1999). Suppression of T and B lymphocyte activation by a *Yersinia pseudotuberculosis* virulence factor, yopH. *J. Exp. Med.* **190**, 1343–50.

Zeidner, N., Mbow, M. L., Dolan, M., Massung, R., Baca, E., and Piesman, J. (1997). Effects of *Ixodes scapularis* and *Borrelia burgdorferi* on modulation of the host immune response: induction of a TH2 cytokine response in Lyme disease-susceptible (C3H/HeJ) mice but not in disease-resistant (BALB/c) mice. *Infect. Immun.* **65**, 3100–6.

Zuany-Amorim, C., Sawicka, E., Manlius, C., Le Moine, A., Brunet, L. R., Kemeny, D. M., Bowen, G., Rook, G., and Walker, C. (2002). Suppression of airway eosinophilia by killed *Mycobacterium vaccae*-induced allergen-specific regulatory T-cells. *Nat. Med.* **8**, 625–9.

CHAPTER 11

Dendritic cells in the gut and their possible role in disease

Christoph Becker
University of Mainz

The gut represents the largest lymphoid tissue of the whole body. The delicate task of the intestinal immune system is the discrimination of harmless food antigens and the commensal bacterial flora from harmful pathogens. Under normal physiologic conditions, immune tolerance is induced to non-pathogenic stimuli while effective immune responses are generated toward dangerous pathogens. Thus "decision making" is an important feature of the intestinal immune system. If inappropriate responses are generated, serious inflammation of the small and large intestine may develop. Crohn's disease (CD) and ulcerative colitis are the two prototypes of such inflammatory bowel disease that are believed to develop as a consequence of a disregulated immune response toward harmless antigens. Despite our limited knowledge on the mechanisms of such "decision making" in the gut, recent evidence suggest an important role of intestinal dendritic cells[1]. Dendritic cells (DCs) can be found in large numbers throughout the gastrointestinal tract where they usually build a tight network underlying the epithelium[1,2]. This chapter will discuss their contribution to the induction of tolerance and immunity in the intestinal immune system as well as a possible role of these DCs in localized immune responses predisposing the terminal ileum for the development of inflammatory bowel disease (IBD).

11.1 DENDRITIC CELLS IN THE INTESTINAL IMMUNE SYSTEM: AN OVERVIEW

The intestinal immune system can be functionally separated into an inductive site and an effector site[1]. The prototypic inductive site in the small intestine is the Peyer's patch, a localized lymphoid structure placed within the bowel wall. Peyer's patches are composed of a specialized, so-called follicle-associated epithelium, the adjacent so-called subepithelial dome, B cell follicles containing germinal centers and surrounding interfollicular regions. Antigens and microorganisms are thought to be transported from the gut lumen via M-cells, specialized epithelial cells within the follicle-associated epithelium, into the Peyer's patch follicle. In the subepithelial dome region they get taken up by local dendritic cells. Dendritic cells in the subepithelial dome represent an immature phenotype. However, upon antigen encounter, they are thought to mature and migrate into the T cell rich area of the interfollicular regions. Here the dendritic cells display a mature phenotype with high expression of the costimulatory molecules CD80 and CD86 thus representing potent antigen-presenting cells, ready to stimulate resident naïve T cells. Stimulated T cells can then circulate and migrate into the lamina propria which is the main effector site of the intestinal immune system.

Dendritic cells in the Peyer's patches like elsewhere in the body commonly express the integrin CD11c. However, three distinct subpopulations of dendritic cells have been identified in murine Peyer's patches based on their expression of the molecules CD11b and CD8α: CD11c$^+$ CD11b$^+$ CD8α^- myeloid dendritic cells, CD11c$^+$ CD11b$^-$ CD8α^+ lymphoid dendritic cells, CD11c$^+$ CD11b$^-$ CD8α^- double negative (DN) dendritic cells[3]. Interestingly these subpopulations are not evenly distributed within the Peyer's patch; they rather reside in distinct areas[3,4]. While the myeloid CD11b$^+$ DCs were found almost exclusively in the subepithelial dome, lymphoid CD8α^+ DCs were found in the interfollicular regions. Third, double negative DCs were found at both locations and in addition within the follicle-associated epithelium[3,5]. Despite the fact that most studies on the role of DCs in the gut were performed with cells isolated from Peyer's patches or mesenteric lymph nodes, DCs are also prominent in the mucosa[1]. High numbers of DCs form a close meshed network within the lamina propria of the small intestine, especially within the villi. In contrast to the small intestine, the colon has no Peyer's patches, but houses numerous isolated lymphoid follicles with some structural similarities to Peyer's patches implicating that they are also involved in inducing immune responses. In addition aggregates of DCs can also be found in the lamina propria of the colon, although at lower numbers than in the small intestine.

11.1.1 Tolerance induction by intestinal dendritic cells

Although the intestinal immune system has to respond to pathogenic infections with the induction of an appropriate cell-mediated immune response, the primary task of the mucosal immune system is the induction of tolerance. Their strategic localization directly underneath the intestinal epithelium or the follicle-associated epithelium enables DCs to immediately respond to the uptake of antigens. Our knowledge of the mechanisms of tolerance induction is very limited, but there is compelling evidence for a tolerizing function of intestinal DCs[5–8]. Accordingly it has been shown, that DCs from Peyer's patches or the lamina propria have a tendency to induce T helper cell 2 (TH2) type immune responses and to express anti-inflammatory cytokines like IL-10 and TGF-β[5,8–10]. Such TH2-like cells express IL-6 which together with TGF-β can promote the production of immunoglobulin A (IgA) by B cells in the follicles of Peyer's patches[11]. More recent data suggests that epithelial cells can educate DCs to induce such TH2-like immune responses by T cells[8]. Thymic stromal lymphopoietin was found to be constitutively released by epithelial cells leading to a conditioning of colon DCs to induce TH2 rather than TH1 immune responses. A second mechanism by which mucosal DCs could induce tolerance is the induction of regulatory T cells (Tregs). It has been demonstrated, that plasmacytoid DCs (CD8α^+ CD11c^{LO}) can induce IL-10 producing regulatory T cells[7]. In another study, antigen-specific CD4$^+$ CD25$^+$ regulatory T cells were found to be induced in Peyer's patches upon oral administration of protein antigen[12,13]. Recently, TGF-β was found to induce regulatory T cells from naïve CD4$^+$ CD25$^-$ cells in the periphery[14,15]. TGF-β seems to act via an induction of FOXP3, the master-regulator of regulatory T cell commitment in naturally occurring (thymus derived) Treg. Since TGF-β is highly expressed in the Peyer's patches within 6 h after oral administration of antigen[16], it is tempting to speculate that such TGF-β released from cells including DCs in Peyer's patches induces regulatory T cells which then recirculate and migrate to the mucosal effector sites in order to induce tolerance against this antigen.

11.1.2 Routes of antigen uptake in the mucosal immune system

To participate in the induction of tolerance and immunity in the gut, lamina propria dendritic cells have to get into contact with antigens present in the lumen of the gut. As mentioned above, M cells located in the follicle-associated epithelium of Peyer's patches are thought to be the most important sites for the sampling of antigens from the gut lumen. However, recent evidence question the necessity of Peyer's patches and support a role for alternative gateways. Accordingly, cells with features of M cells of Peyer's patches have been described in the (non-follicle-associated) epithelial cell layer of intestinal villi[17]. Interestingly, such villous M cells were found not only in wildtype mice, but also in mice genetically deficient for Peyer's patches, indicating that they develop independent from Peyer's patches. Functionally it was shown that villous M cells were capable of translocating bacteria like *Salmonella* and *Yersinia*[17]. Dendritic cells can also directly sample bacteria from the gut lumen: it has been impressively demonstrated that lamina propria DCs can extend their dendrites through the epithelial cell layer into the lumen in order to directly sample bacterial antigens out of the lumen[18]. Interestingly these DCs express tight junction proteins like occludin, claudin-1 and zonula occludens-1 and can thereby prevent the disruption of the epithelial barrier. The capability of DCs to directly access the lumen and sample bacteria may provide an important function in the probing for the presence of pathogenic microorganisms in the gut lumen. To date it is not clear whether DCs extend their dendrites constitutively or whether this is an induced mechanism in response to certain bacteria. The observation that such dendrite forming DCs are more prominent in the terminal ileum than in more proximal regions may support the latter hypothesis, given that there is a gradient of bacteria toward the terminal ileum of the small intestine[19]. The formation of such transepithelial dendrites by lamina propria DCs was found to depend on the chemokine receptor CX3CR1, also called *fractalkine*[19]. CX3CR1 was found to be important for the clearance of gut-invasive pathogens by DCs.

An additional route of antigen uptake in the intestine may be the transport of apoptotic epithelial cells to the mesenteric lymph nodes by lamina propria DCs which occurs in the steady-state[20]. Such a continuous circulation of DCs from the lamina propria to the lymphatic system has been proposed as a mechanism by which tolerance to normal antigens in the gut is conferred. The constitutive migration of DCs from the intestinal mucosa to lymphatic vessels could be experimentally accelerated by a systemic application of inflammatory cytokines or bacterial products[21,22]. The underlying mechanism leading to such an induced and guided movement of DCs in the gut may be the coordinated regulation of the expression of chemokines and their receptors.

11.2 DENDRITIC CELLS IN THE PATHOGENESIS OF INFLAMMATORY BOWEL DISEASE

Inflammatory bowel disease (IBD) comprises Crohn's disease and ulcerative colitis, the two major chronic inflammatory diseases of the intestine. The cause of IBD is believed to be multifactorial[23]. It seems that environmental as well as genetic factors in combination with the microbial intestinal flora and a disregulated immune response trigger the development of chronic inflammation of the intestine. As described above, the immune system of the gastrointestinal tract is sensitively balanced. The processes that lead to a loss of this balance are poorly understood. If however tolerance against food antigens or the commensal flora is lost, a deleterious process is evoked, leading to a disregulated activation of mucosal immune cells such as DCs and T cells.

11.2.1 Dendritic cells play a major role in the pathogenesis of colitis

Increasing evidence suggests that intestinal bacterial flora is involved in the pathogenesis of IBD[24,25]. Despite considerable effort, to date no specific bacterial pathogen has been identified, implicating that different bacteria are involved in the initiation of the pathogenic immune response[26,27]. Recent data suggest that DCs in the intestine are the link between the bacterial flora and the persistent inflammation in the colon as seen in patients with IBD, implicating that a disregulated handling of such bacteria play a major role in the early events that lead to the development of chronic inflammatory and autoimmune diseases. Recently, it was found that a subset of Crohn's disease (CD) patients have mutations in the gene that encodes the nucleotide-binding oligomerization domain 2 (*NOD2*) protein[28]. *NOD2* has been described as a pathogen-recognition receptor that recognizes muramyl dipeptide derived from bacterial peptidoglycans[29]. In another study, NOD2 deficiency in mice or the presence of a CD-like NOD2 mutation was shown to increase Toll-like receptor 2-mediated activation of NF-kappaB and also TH1 responses[30]. Thus, NOD2 mutations may lead to disease by causing excessive TH1 responses, suggesting that alterations in the recognition of intestinal bacteria can contribute to IBD.

Numerous mouse models of IBD have been developed. From all these models, the so-called adoptive transfer model has drawn much attention in recent times[31,32]. This model relies on the adoptive transfer of colitogenic

CD4$^+$CD45RB^{HI} T cells into SCID recipient mice. SCID mice have a mutant phenotype and as a result of this do not develop functional T cells and B cells. The transferred colitogenic T cells repopulate the SCID hosts and lead to a severe chronic colitis after 8 to 12 weeks. Importantly the colon pathology very much resembles that seen in patients with CD, making it an important model for the investigation of human IBD[31]. Studies performed with this model implicated an important role for DCs in the early events of colitis manifestation. Transferred T cells were demonstrated to form aggregates with CD11c$^+$ DC in the lamina propria, where they substantially proliferated[33]. Interestingly, the authors of this report observed a delay between the influx of T cells into the mesenteric lymph nodes and the formation of aggregates between T cells and DCs in the lamina propria. The authors hypothesized that after adoptive transfer, T cells went to the mesenteric lymph nodes to be primed by DCs before they finally migrated to the lamina propria to fulfill their effector function. Other studies also demonstrated that a priming of the T cells by DCs was necessary for the development of colitis in the adoptive transfer model[34]. In this study, the priming of colitogenic T cells was shown to be dependent on OX40L since blocking the interaction of OX40L present on T cells with OX40 present on DCs protected mice from colitis development. Consequently high numbers of OX40L expressing CD11c$^+$ CD11b$^+$ CD8α^- DCs were found in mice with colitis. Thus it seems that after adoptive transfer but before the onset of disease, T cells are primed by a direct interaction with DCs in the mesenteric lymph nodes and subsequently migrate into the lamina propria where they undergo a massive proliferation which then finally leads to a chronic colitis characterized by the production of vast amount of inflammatory cytokines. Not only do T cells proliferate in the course of disease development, but also the expansion of DCs in the lamina propria of colitis bearing animals has been described[35]. These DCs were shown to display an upregulation of costimulatory molecules like CD80, CD86 and CD40. Therefore, it seems that T cells and DCs in the lamina propria support each other's growth and stimulation, thereby sustaining the inflammation.

11.2.2 Dendritic cells polarize T helper cell responses leading to Crohn's disease

Despite a general role of DCs in the priming of T cells and the induction of colitis, DCs could also specifically direct the kind of T helper cell response generated in IBD patients. Recent evidence suggest that in CD, T cells represent a dominant TH1 phenotype while in ulcerative colitis T cells are

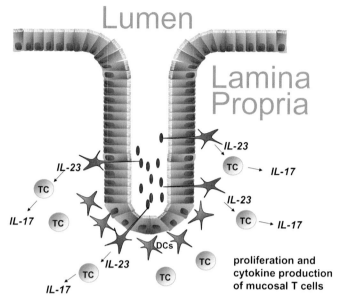

Lumen

Lamina
Propria

IL-23

IL-23

TC → *IL-17*

TC

IL-17 TC

IL-23

TC → *IL-17*

DCs

IL-23

TC

TC

proliferation and
cytokine production
of mucosal T cells

IL-17

Figure 11.1. Model of constitutive IL-23 production in the terminal ileum: dendritic cells (DCs) sample bacteria from the lumen and produce IL-23. IL-23 in turn can stimulate infiltrating memory T cells to proliferate and produce cytokines, including IL-17.

predominantly skewed toward a TH2-like phenotype[36]. When CD patients were compared to control patients, isolated lamina propria CD4$^+$ T cells from CD samples produced elevated levels of the TH1 cytokine IFN-γ while the expression of the TH2 cytokines IL-4 and IL-5 were diminished. In line with such a polarized model, IL-12 produced by activated DCs has been proposed as a key cytokine in the pathogenesis of CD but not in the pathogenesis of ulcerative colitis[37–39]. IL-12 (p40/p35) is rapidly released from DCs after stimulation via Toll-like receptors and studies in mouse models of CD have reported an essential role of IL-12 in the pathogenesis of chronic intestinal inflammation. Neutralizing IL-12 p40 in these animals using a specific antibody resulted in the protection from IBD[39–41]. Further studies revealed that mice transgenic for STAT-4, a signaling molecule downstream of the IL-12 receptor, are highly suscep-tible to TH1 driven colitis[42]. Antibodies against IL-12 p40 are currently in clinical trials. A recent phase-2 study reported long-lasting remission in 50 per cent of patients treated[43]. Thus IL-12 is a promising target for future therapeutic intervention in patients with CD. However, many of the investigations regarding IL-12 were performed using antibodies against the IL-12 p40 subunit.

Since the discovery that in DCs IL-12 p40 can also form heterodimers with the recently discovered molecule p19 resulting in IL-23[44], these data have to be interpreted with care. Studies using IL-23 (p19) knockout mice in colitis are still missing, but the discovery that in collagen-induced arthritis and in experimental autoimmune encephalomyelitis, two commonly used models of chronic inflammation and autoimmunity, IL-23 knockout mice but not IL-12 knockout mice are protected from disease, implicate an important role for DC derived IL-23 also in intestinal immune pathology[45,46], Support for a role of IL-23 in IBD comes from a recent study using mice transgenic for IL-23. Transgenic mice constitutively overexpressing IL-23 p19 developed a spontaneous severe inflammation in multiple organs including the gut, with elevated levels of proinflammatory cytokines[47]. In another study, it was demonstrated that IL-23 p19 was significantly increased in the inflamed mucosa of patients with CD and to a lesser extent in patients with ulcerative colitis as compared to patients without IBD[48,49]. Moreover in these studies, elevation of IL-23 p19 levels in CD correlated with the severity of the lesions observed during endoscopy. In a similar study, IL-17, a cytokine induced in T cells upon stimulation with IL-23, was found to be highly expressed in the serum and in the mucosa of patients with IBD, but not in control patients[50,51]. Interestingly IL-17 was increased not only in CD patients but also in ulcerative colitis patients, implicating an alternative route of IL-17 induction not dependent on IL-23 in ulcerative colitis pathogenesis. Given what is known from studies in other model systems, DC derived IL-12 may be an important pathogenic factor in the onset of IBD, while later on, DC expressed IL-23 may drive strong memory T cell responses leading to the perpetuation of IBD. In line with this model, IL-23 has been shown to stimulate naïve murine T cells only poorly, an observation that is resembled by the low expression of IL-23R on the surface of the latter cells[52]. In contrast, activated memory T cells respond strongly to IL-23 but only poorly to IL-12, a finding that is paralleled by the high expression of the IL-23R on the surface of memory T cells[45,46]. IL-23 has therefore been proposed to be a late stage effector cytokine rather than a cytokine inducing T cell commitment into a certain lineage.

However, this concept is challenged by recent reports, which have demonstrated that IL-23 stimulation can lead to the generation of an alternative T helper cell subset characterized by the expression of high levels of the proinflammatory cytokine IL-17 but only low amounts of IFN-γ[46,53–56]. This novel T cell population was denoted TH_{IL-17} and is currently under investigation. A recent report by Langrish and colleagues highlights the role of such TH_{IL-17} cells in an animal model of EAE[56].

In this report the authors demonstrate that IL-23 dependent TH_{IL-17} cells drive autoimmune inflammation in the brain and that neutralization of soluble IL-17 by using antibodies partially protected mice from EAE. In the animal model of collagen-induced arthritis, resistance of IL-23 knockout mice was found to depend on the absence of IL-17 producing T helper cells but not on an impaired TH1 immune response, highlighting again the possible role of TH_{IL-17} cells for chronic and autoimmune inflammation[46]. Both studies demonstrated higher disease levels in IL-12 (p35) knockout mice when compared to wild-type mice. The authors attributed this observation to the higher level of IL-17 producing T helper cells in the absence of IFN-γ producing cells. Thus IL-23 induced IL-17 expression emerges as an important pathway in the pathogenesis of chronic inflammatory and autoimmune diseases including inflammatory bowel disease. IL-17 has furthermore been described as a cytokine that drives proliferation, maturation and chemotaxis of neutrophils[57]. Beyond its role on neutrophils, IL-17 has been shown to induce the expression of other proinflammatory cytokines like TNF-α, IL-1 and IL-6 in various cell types[58–60]. IL-17 has also been demonstrated to drive the maturation of dendritic cells[61].

However to date it is unclear whether IL-23 alone or in combination with other stimuli can induce the generation of IL-17 producing T cells directly from naïve T cells or whether they represent a population that has been generated from previously activated T helper cells or early activated T helper cells before terminal commitment. Another possibility would be that these cells develop from a subset of TH1 cells. In support of the latter theory, a recent report by Wu and colleagues has demonstrated the existence of a novel IFN-γ negative TH1 cell population[62]. In their model, IFN-γ producing TH1 cells showed immediate effector function and were short lived, while the alternative IFN-γ negative population was long-lived and displayed memory function. It remains to be investigated whether the cell population described in this paper resembles cells that can develop toward the TH_{IL-17} phenotype.

This view is also supported by studies demonstrating that IL-12 (p40) is sustaining chronic inflammation in the colon of IL-10 knockout mice while IFN-γ is dispensable[41]. While other studies reported a crucial role of IFN-γ for the pathogenesis of IBD[63,64], it was shown in an adoptive transfer model of colitis, that $CD45RB^{Hi}$ cells from IFN-γ knockout mice were still able to induce colitis, although with less severity and with a later onset than wild-type $CD45RB^{Hi}$ cells[65]. Thus it seems that IBD can develop independently from IFN-γ. Furthermore, antibodies neutralizing

IFN-γ were only able to suppress colitis when given before the onset of colitis but not in established colitis[66]. Thus although IFN-γ is important in the initial events leading to IBD, it may be dispensable in established colitis.

The third member of IL-12 related cytokines produced by activated DCs is IL-27. IL-27 is a heterodimer composed of the two subunits EBI3 and p28[67]. IL-27 has been shown to cooperate with IL-12 in the priming of naïve T cells and has also been shown to play an important role in the pathogenesis of IBD[68–70].

In summary, in CD, activated DCs in the gut seem to drive inflammation through the release of high levels of inflammatory cytokines, especially IL-12 and the IL-12 related cytokines IL-23 and IL-27. It has to be mentioned that beside IL-12 related cytokines, other DC derived cytokines like IL-6 and TNF-α have been shown to play an important role in the pathogenesis, especially in the perpetuation of IBD as well[71,72]. These findings implicate that in the onset of disease, DC expressed IL-12 and IL-27 may drive the generation of a TH1 response while in later stages the disease is perpetuated by other DC derived cytokines like IL-23, IL-6 and TNF-α. Therefore, the important role of the above mentioned cytokines highlights the role of DCs in the pathogenesis of IBD. In the case of TNF-α, monoclonal antibodies targeting this cytokine have even made their way from bench to bedside and the beneficial use of this antibody in patients with CD can be seen as a success story of targeting dendritic cells and their mediators in disease.

11.3 A ROLE FOR LOCAL DENDRITIC CELLS IN THE PREDISPOSITION OF THE TERMINAL ILEUM FOR COLITIS

Inflammatory bowel diseases do not affect all segments of the intestine evenly[25]. Crohn's disease may exclusively affect the small bowel, especially the terminal ileum, which is the most common location, or it may involve the small bowel and the colon. In contrast, in some patients the involvement is limited to the colon. In more rare cases even other segments of the gastrointestinal tract like the stomach, duodenum or the esophagus are affected by CD. Ulcerative colitis most frequently develops in the distal colon and progressively involves the whole colon up to the ileal junction. However, ulcerative colitis is limited to the colon and does not affect the small intestine. Experimental models for IBD in the mouse most frequently affect the colon and histologic signs of colitis are most severe in the distal colon[73]. In the mouse system only few models exist in which the small intestine is involved. However, it has to be mentioned that some frequently used experimental colitis models are based on the rectal administration of reagents like TNBS, oxazolone or ethanol which induce inflammation only in the contact area of the colon.

The observation that in mice and men distinct regions of the small and large intestine are more or less prone to develop inflammation implicates that the small and large intestine immunologically cannot be considered a homogenous organ. The anatomical architecture in different parts of the intestine may directly reflect its functional role in digestion and host defense[1]. In agreement with such a model, differences have been described not only in the intestinal epithelial architecture, but also in the distribution and density of cells that are associated with the innate and adaptive immune response in the gut, such as DCs, T and B lymphocytes and Paneth cells. In addition, lymphoid structures like Peyer's patches or isolated lymphoid follicles are not evenly distributed throughout the intestine. Since different strains of bacteria colonize distinct regions of the intestine and since there are differences in bacterial concentrations throughout the gut[24], it is likely that the lymphoid tissue throughout the intestine is highly adapted to the environment and plays an important role in mediating localized immune responses to bacteria that populate the respective compartments. The development of such a localized immune system may therefore only be in part genetically determined. In fact, recent evidence suggests that it is functionally dependent on the presence of the bacterial microflora[74]. Interestingly, in HLA-B27 transgenic rats, the severity of mucosal inflammation is determined by the bacterial load in the cecum[75,76]. The development of colitis in T cell receptor mutant mice can be prevented by removal of parts of the cecum[77]. In addition, differential complementary determining region-3 T cell receptor usage among T cells in different regions of the colon provides further support for localized immune responses within the large intestine[78]. Recently it was demonstrated that antigen processing by 20S proteasomes is more effective in the small bowel as compared to the colon and other organs[79]. These results imply that the manipulation of bacteria or the corresponding mucosa-associated lymphoid tissue in specific compartments of the gut can influence the induction of intestinal immunopathology.

Further evidence for localized immune responses in the gut comes from a study which recently demonstrated that dendritic cells in the lamina propria of the terminal ileum are constitutively activated by luminal bacterial antigens, giving rise to a constitutive presence of interleukin-23 only in this part of the murine gut[80]. In this study, a subset of lamina propria DCs was identified displaying constitutive IL-12 p40 promoter activity and IL-23 expression in the terminal ileum. Constitutively activated LPDC in this study were demonstrated to be largely CD8α and CD11b double-negative DCs and thus reminiscent of a DC subset recently characterized in murine Peyer's patches that produces high levels of IL-12 in response to *Staphylococcus aureus* antigen and IFN-gamma stimulation[5].

The above study raised the question about the cause of a localized response as indicated by the difference in IL-12 p40 and IL-23 production between LPDC from the proximal and distal small intestine of healthy mice. One possible explanation may be that LPDC could be differentially activated through the local microenvironment. In agreement with this hypothesis, the authors demonstrated localization of bacteria directly in p40 expressing cells and endocytosis of bacteria by such LPDC in the crypts of the terminal ileum but not the proximal parts of the small bowel. These differences might be due to an increased bacterial load of the terminal ileum as compared to the proximal small bowel or to local changes in the composition of the microflora. It is known that there is a general gradient of bacterial concentration increasing toward the terminal ileum where in CD the highest incidence of lesions can be found. Consistent with a dominant role for resident bacteria, no increase in constitutive p40 expression was observed in the terminal ileum of mice raised under germ-free conditions. Thus, this study implicates important functional differences between the mucosal immune systems of the colon and the proximal and distal small intestine and provided evidence for localized DC driven immune responses in the gut. Interestingly, an uptake of bacteria by LPDC of the terminal ileum was only seen in the crypts where the earliest pathological manifestations of ileitis in CD are known to occur. Based on the above findings, it is tempting to speculate, that the elevation of IL-23 expression selectively in the ileum reflects an increased susceptibility to inflammation in the terminal ileum, a site that is most frequently affected in patients with CD[25].

Since the functional IL-23 receptor is expressed mainly on memory T cells and T cells in the intestinal mucosa largely represent a memory phenotype, memory T cells migrating into the terminal ileum may be highly susceptible to locally elevated levels of IL-23. Therefore infiltrating T cells may develop toward inflammatory TH$_{IL-17}$ cells promoting inflammation in the terminal ileum but not in other segments of the intestine. Despite these findings, localized innate and adaptive mucosal immune responses may also be protective for the intestine, providing an efficient response to the respective flora while restricting extensive immunopathology.

REFERENCES

1. Mowat, A. M. (2003). Anatomical basis of tolerance and immunity to intestinal antigens. *Nat. Rev. Immunol.* **3**, 331–41.

2. Bilsborough, J. and Viney, J. L. (2004). Gastrointestinal dendritic cells play a role in immunity, tolerance, and disease. *Gastroenterology* **127**, 300–9.

3. Iwasaki, A. and Kelsall, B. L. (2000). Localization of distinct Peyer's patch dendritic cell subsets and their recruitment by chemokines macrophage inflammatory protein (Mip)-3alpha, Mip-3beta, and secondary lymphoid organ chemokine. *J. Exp. Med.* **191**, 1381–94.

4. Kelsall, B. L. and Strober, W. (1996). Distinct populations of dendritic cells are present in the subepithelial dome and T cell regions of the murine Peyer's patch. *J. Exp. Med.* **183**, 237–47.

5. Iwasaki, A., and Kelsall, B. L. (2001). Unique functions of Cd11b$^+$, Cd8 alpha$^+$, and double-negative Peyer's patch dendritic cells. *J. Immunol.* **166**, 4884–90.

6. Macpherson, A. J and Uhr, T. (2004). Induction of protective Iga by intestinal dendritic cells carrying commensal bacteria. *Science* **303**, 1662–5.

7. Bilsborough, J., George, T. C., Norment, A., and Viney, J. L. (2003). Mucosal Cd8alpha$^+$ Dc, with a plasmacytoid phenotype, induce differentiation and support function of T cells with regulatory properties. *Immunology* **108**, 481–92.

8. Rimoldi, M., Chieppa, M., Salucci, V., Avogadri, F., Sonzogni, A., Sampietro, G. M., Nespoli, A., Viale, G., Allavena, P., and Rescigno, M. (2005). Intestinal immune homeostasis is regulated by the crosstalk between epithelial cells and dendritic cells. *Nat. Immunol.* **6**, 507–14.

9. Akbari, O., DeKruyff, R. H., and Umetsu, D. T. (2001). Pulmonary dendritic cells producing Il-10 mediate tolerance induced by respiratory exposure to antigen. *Nat. Immunol.* **2**, 725–31.

10. Alpan, O., Rudomen, G., and Matzinger, P. (2001). The role of dendritic cells, B cells, and M cells in gut-oriented immune responses. *J. Immunol.* **166**, 4843–52.

11. Sato, A., Hashiguchi, M., Toda, E., Iwasaki, A., Hachimura, S., and Kaminogawa, S. (2003). Cd11b$^+$ Peyer's patch dendritic cells secrete Il-6 and induce Iga secretion from naive B cells. *J. Immunol.* **171**, 3684–90.

12. Tsuji, N. M., Mizumachi, K., and Kurisaki, J. (2003). Antigen-specific, Cd4$^+$Cd25$^+$ regulatory T cell clones induced in Peyer's patches. *Int. Immunol.* **15**, 525–34.

13. Hauet-Broere, F., Unger, W. W., Garssen, J., Hoijer, M. A., Kraal, G., and Samsom, J. N. (2003). Functional Cd25$^-$ and Cd25$^+$ mucosal regulatory

T cells are induced in gut-draining lymphoid tissue within 48 h after oral antigen application. *Eur. J. Immunol.* **33**, 2801–10.

14. Chen, W., Jin, W., Hardegen, N., Lei, K. J., Li, L., Marinos, N., McGrady, G., and Wahl, S. M. (2003). Conversion of peripheral Cd4$^+$Cd25$^-$ naive T cells to Cd4$^+$Cd25$^+$ regulatory T cells by Tgf-beta induction of transcription factor Foxp3. *J. Exp. Med.* **198**, 1875–86.

15. Fantini, M. C., Becker, C., Monteleone, G., Pallone, F., Galle, P. R, and Neurath, M. F. (2004). Cutting edge: Tgf-beta induces a regulatory phenotype in Cd4$^+$Cd25$^-$ T cells through Foxp3 induction and down-regulation of Smad7. *J. Immunol.* **172**, 5149–53.

16. Gonnella, P. A., Chen, Y., Inobe, J., Komagata, Y., Quartulli, M., and Weiner, H. L. (1998). In situ immune response in gut-associated lymphoid tissue (Galt) following oral antigen in Tcr-transgenic mice. *J. Immunol.* **160**, 4708–18.

17. Jang, M. H., Kweon, M. N., Iwatani, K., Yamamoto, M., Terahara, K., Sasakawa, C., Suzuki, T., Nochi, T., Yokota, Y., Rennert, P. D., Hiroi, T., Tamagawa, H., Iijima, H., Kunisawa, J., Yuki, Y., and Kiyono, H. (2004). Intestinal villous M cells: an antigen entry site in the mucosal epithelium. *Proc. Natl Acad. Sci. U S A* **101**, 6110–15.

18. Rescigno, M., Urbano, M., Valzasina, B., Francolini, M., Rotta, G., Bonasio, R., Granucci, F., Kraehenbuhl, J. P., and Ricciardi-Castagnoli, P. (2001). Dendritic cells express tight junction proteins and penetrate gut epithelial monolayers to sample bacteria. *Nat. Immunol.* **2**, 361–7.

19. Niess, J. H., Brand, S., Gu, X., Landsman, L., Jung, S., McCormick, B. A., Vyas, J. M., Boes, M., Ploegh, H. L., Fox, J. G., Littman, D. R., and Reinecker, H. C. (2005). Cx3cr1-Mediated dendritic cell access to the intestinal lumen and bacterial clearance. *Science* **307**, 254–8.

20. Huang, F. P., Platt, N., Wykes, M., Major, J. R., Powell, T. J., Jenkins, C. D., and MacPherson, G. G. (2000). A discrete subpopulation of dendritic cells transports apoptotic intestinal epithelial cells to T cell areas of mesenteric lymph nodes. *J. Exp. Med.* **191**, 435–44.

21. MacPherson, G. G., Jenkins, C. D., Stein, M. J., and Edwards, C. (1995). Endotoxin-mediated dendritic cell release from the intestine. Characterization of released dendritic cells and Tnf dependence. *J. Immunol.* **154**, 1317–22.

22. Roake, J. A., Rao, A. S., Morris, P. J., Larsen, C. P., Hankins, D. F., and Austyn, J. M. (1995). Dendritic cell loss from nonlymphoid tissues after systemic administration of lipopolysaccharide, tumor necrosis factor, and interleukin 1. *J. Exp. Med.* **181**, 2237–47.

23. Bouma, G. and Strober, W. (2003). The immunological and genetic basis of inflammatory bowel disease. *Nat. Rev. Immunol.* **3**, 521–33.

24. Guarner, F. and Malagelada, J. R. (2003). Gut flora in health and disease. *Lancet* **361**, 512–19.

25. Shanahan, F. (2002). Crohn's disease. *Lancet* **359**, 62–9.

26. Swidsinski, A., Ladhoff, A., Pernthaler, A., Swidsinski, S., Loening-Baucke, V., Ortner, M., Weber, J., Hoffmann, U., Schreiber, S., Dietel, M., and Lochs, H. (2002). Mucosal flora in inflammatory bowel disease. *Gastroenterology* **122**, 44–54.

27. Linskens, R. K., Huijsdens, X. W., Savelkoul, P. H., Vandenbroucke-Grauls, C. M., and Meuwissen, S. G. (2001). The bacterial flora in inflammatory bowel disease: current insights in pathogenesis and the influence of antibiotics and probiotics. *Scand. J. Gastroenterol. Suppl.* 29–40.

28. Ogura, Y., Bonen, D. K., Inohara, N., Nicolae, D. L., Chen, F. F., Ramos, R., Britton, H., Moran, T., Karaliuskas, R., Duerr, R. H., Achkar, J. P., Brant, S. R., Bayless, T. M., Kirschner, B. S., Hanauer, S. B., Nunez, G., and Cho, J. H. (2001). A frame shift mutation in Nod2 associated with susceptibility to Crohn's disease. *Nature* **411**, 603–6.

29. Inohara, N., Ogura, Y., Fontalba, A., Gutierrez, O., Pons, F., Crespo, J., Fukase, K., Inamura, S., Kusumoto, S., Hashimoto, M., Foster, S. J., Moran, A. P., Fernandez-Luna, J. L., and Nunez, G. (2003). Host recognition of bacterial muramyl dipeptide mediated through Nod2. Implications for Crohn's disease. *J. Biol. Chem.* **278**, 5509–12.

30. Watanabe, T., Kitani, A., Murray, P. J., and Strober, W. (2004). Nod2 is a negative regulator of Toll-like receptor 2-mediated T helper type 1 responses. *Nat. Immunol.* **5**, 800–8.

31. Leach, M. W., Bean, A. G., Mauze, S., Coffman, R. L., and Powrie, F. (1996). Inflammatory bowel disease in C.B-17 Scid mice reconstituted with the Cd45rbhigh subset of Cd4[+] T cells. *Am. J. Pathol.* **148**, 1503–15.

32. Powrie, F., Leach, M. W., Mauze, S., Caddle, L. B., and Coffman, R. L. (1993). Phenotypically distinct subsets of Cd4[+] T cells induce or protect from chronic intestinal inflammation in C.B-17 Scid mice. *Int. Immunol.* **5**, 1461–71.

33. Leithauser, F., Trobonjaca, Z., Moller, P., and Reimann, J. (2001). Clustering of colonic lamina propria Cd4(+) T cells to subepithelial dendritic cell aggregates precedes the development of colitis in a murine adoptive transfer model. *Lab. Invest.* **81**, 1339–49.

34. Malmstrom, V., Shipton, D., Singh, B., Al-Shamkhani, A., Puklavec, M. J., Barclay, A. N., and Powrie, F. (2001). Cd134l expression on dendritic cells

in the mesenteric lymph nodes drives colitis in T cell-restored Scid mice. *J. Immunol.* **166**, 6972–81.

35. Krajina, T., Leithauser, F., Moller, P., Trobonjaca, Z., and Reimann, J. (2003). Colonic lamina propria dendritic cells in mice with Cd4$^+$ T cell-induced colitis. *Eur. J. Immunol.* **33**, 1073–83.

36. Fuss, I. J., Neurath, M., Boirivant, M., Klein, J. S., de la Motte, C., Strong, S. A., Fiocchi, C., and Strober, W. (1996). Disparate Cd4$^+$ lamina propria (Lp) lymphokine secretion profiles in inflammatory bowel disease. Crohn's disease Lp cells manifest increased secretion of Ifn-gamma, whereas ulcerative colitis Lp cells manifest increased secretion of Il-5. *J. Immunol.* **157**, 1261–70.

37. Stuber, E., Strober, W., and Neurath, M. (1996). Blocking the Cd40l–Cd40 interaction in vivo specifically prevents the priming of T helper 1 cells through the inhibition of interleukin 12 secretion. *J. Exp. Med.* **183**, 693–8.

38. Monteleone, G., Biancone, L., Marasco, R., Morrone, G., Marasco, O., Luzza, F., and Pallone, F. (1997). Interleukin 12 is expressed and actively released by Crohn's disease intestinal lamina propria mononuclear cells. *Gastroenterology* **112**, 1169–78.

39. Neurath, M. F., Fuss, I., Kelsall, B. L., Stuber, E., and Strober, W. (1995). Antibodies to interleukin 12 abrogate established experimental colitis in mice. *J. Exp. Med.* **182**, 1281–90.

40. Simpson, S. J., Shah, S., Comiskey, M., de Jong, Y. P., Wang, B., Mizoguchi, E., Bhan, A. K., and Terhorst, C. T. (1998). T cell-mediated pathology in two models of experimental colitis depends predominantly on the interleukin 12/signal transducer and activator of transcription (Stat)-4 pathway, but is not conditional on interferon gamma expression by T cells. *J. Exp. Med.* **187**, 1225–34.

41. Davidson, N. J., Hudak, S. A., Lesley, R. E., Menon, S., Leach, M. W., and Rennick, D. M. (1998). Il-12, but not Ifn-gamma, plays a major role in sustaining the chronic phase of colitis in Il-10-deficient mice. *J. Immunol.* **161**, 3143–9.

42. Wirtz, S., Finotto, S., Kanzler, S., Lohse, A. W., Blessing, M., Lehr, H. A., Galle, P. R., and Neurath, M. F. (1999). Cutting edge: chronic intestinal inflammation in Stat-4 transgenic mice: characterization of disease and adoptive transfer by Tnf- plus Ifn-gamma-producing Cd4$^+$ T cells that respond to bacterial antigens. *J. Immunol.* **162**, 1884–8.

43. Mannon, P. J., Fuss, I. J., Mayer, L., Elson, C. O., Sandborn, W. J., Present, D., Dolin, B., Goodman, N., Groden, C., Hornung, R. L., Quezado, M., Neurath, M. F., Salfeld, J., Veldman, G. M., Schwertschlag,

U., Strober, W., and Yang, Z. (2004). Anti-interleukin-12 antibody for active Crohn's disease. *N. Engl. J. Med.* **351**, 2069–79.

44. Oppmann, B., Lesley, R., Blom, B., Timans, J. C., Xu, Y., Hunte, B., Vega, F., Yu, N., Wang, J., Singh, K., Zonin, F., Vaisberg, E., Churakova, T., Liu, M., Gorman, D., Wagner, J., Zurawski, S., Liu, Y., Abrams, J. S., Moore, K. W., Rennick, D., de Waal-Malefyt, R., Hannum, C., Bazan, J. F., and Kastelein, R. A. (2000). Novel P19 protein engages Il-12p40 to form a cytokine, Il-23, with biological activities similar as well as distinct from Il-12. *Immunity* **13**, 715–25.

45. Cua, D. J., Sherlock, J., Chen, Y., Murphy, C. A., Joyce, B., Seymour, B., Lucian, L., To, W., Kwan, S., Churakova, T., Zurawski, S., Wiekowski, M., Lira, S. A., Gorman, D., Kastelein, R. A., and Sedgwick, J. D. (2003). Interleukin-23 rather than interleukin-12 is the critical cytokine for autoimmune inflammation of the brain. *Nature* **421**, 744–8.

46. Murphy, C. A., Langrish, C. L., Chen, Y., Blumenschein, W., McClanahan, T., Kastelein, R. A., Sedgwick, J. D., and Cua, D. J. (2003). Divergent pro- and antiinflammatory roles for Il-23 and Il-12 in joint autoimmune inflammation. *J. Exp. Med.* **198**, 1951–7.

47. Wiekowski, M. T., Leach, M. W., Evans, E. W., Sullivan, L., Chen, S. C., Vassileva, G., Bazan, J. F., Gorman, D. M., Kastelein, R. A., Narula, S., and Lira, S. A. (2001). Ubiquitous transgenic expression of the Il-23 subunit P19 induces multiorgan inflammation, runting, infertility, and premature death. *J. Immunol.* **166**, 7563–70.

48. Stallmach, A., Giese, T., Schmidt, C., Ludwig, B., Mueller-Molaian, I., and Meuer, S. C. (2004). Cytokine/chemokine transcript profiles reflect mucosal inflammation in Crohn's disease. *Int. J. Colorectal Dis.* **19**, 308–15.

49. Schmidt, C., Giese, T., Ludwig, B., Mueller-Molaian, I., Marth, T., Zeuzem, S., Meuer, S. C., and Stallmach, A. (2005). Expression of interleukin-12-related cytokine transcripts in inflammatory bowel disease: elevated interleukin-23p19 and interleukin-27p28 in Crohn's disease but not in ulcerative colitis. *Inflamm. Bowel Dis.* **11**, 16–23.

50. Fujino, S., Andoh, A., Bamba, S., Ogawa, A., Hata, K., Araki, Y., Bamba, T., and Fujiyama, Y. (2003). Increased expression of interleukin 17 in inflammatory bowel disease. *Gut* **52**, 65–70.

51. Nielsen, O. H., Kirman, I., Rudiger, N., Hendel, J., and Vainer, B. (2003). Upregulation of interleukin-12 and -17 in active inflammatory bowel disease. *Scand. J. Gastroenterol.* **38**, 180–5.

52. Parham, C., Chirica, M., Timans, J., Vaisberg, E., Travis, M., Cheung, J., Pflanz, S., Zhang, R., Singh, K. P., Vega, F., To, W., Wagner, J., O'Farrell, A. M., McClanahan, T., Zurawski, S., Hannum, C., Gorman, D.,

Rennick, D. M., Kastelein, R. A., de Waal Malefyt, R., and Moore, K. W. (2002). A receptor for the heterodimeric cytokine Il-23 is composed of Il-12rbeta1 and a novel cytokine receptor subunit, Il-23r. *J. Immunol.* **168**, 5699–708.

53. Eijnden, S. V., Goriely, S., De Wit, D., Willems, F., and Goldman, M. (2005). Il-23 up-regulates Il-10 and induces Il-17 synthesis by polyclonally activated naive T cells in human. *Eur. J. Immunol.* **35**, 469–75.

54. Aggarwal, S., Ghilardi, N., Xie, M. H., de Sauvage, F. J., and Gurney, A. L. (2003). Interleukin-23 promotes a distinct Cd4 T cell activation state characterized by the production of interleukin-17. *J. Biol. Chem.* **278**, 1910–14.

55. Happel, K. I., Zheng, M., Young, E., Quinton, L. J., Lockhart, E., Ramsay, A. J., Shellito, J. E., Schurr, J. R., Bagby, G. J., Nelson, S., and Kolls, J. K. (2003). Cutting edge: roles of Toll-like receptor 4 and Il-23 in Il-17 expression in response to *Klebsiella pneumoniae* infection. *J. Immunol.* **170**, 4432–6.

56. Langrish, C. L., Chen, Y., Blumenschein, W. M., Mattson, J., Basham, B., Sedgwick, J. D., McClanahan, T., Kastelein, R. A., and Cua, D. J. (2005). Il-23 drives a pathogenic T cell population that induces autoimmune inflammation. *J. Exp. Med.* **201**, 233–40.

57. Kolls, J. K. and Linden, A. (2004). Interleukin-17 family members and inflammation. *Immunity* **21**, 467–76.

58. Fossiez, F., Djossou, O., Chomarat, P., Flores-Romo, L., Ait-Yahia, S., Maat, C., Pin, J. J., Garrone, P., Garcia, E., Saeland, S., Blanchard, D., Gaillard, C., Das Mahapatra, B., Rouvier, E., Golstein, P., Banchereau, J., and Lebecque, S. (1996). T cell interleukin-17 induces stromal cells to produce proinflammatory and hematopoietic cytokines. *J. Exp. Med.* **183**, 2593–603.

59. Jovanovic, D. V., Di Battista, J. A., Martel-Pelletier, J., Jolicoeur, F. C., He, Y., Zhang, M., Mineau, F., and Pelletier, J. P. (1998). Il-17 stimulates the production and expression of proinflammatory cytokines, Il-beta and Tnf-alpha, by human macrophages. *J. Immunol.* **160**, 3513–21.

60. Molet, S., Hamid, Q., Davoine, F., Nutku, E., Taha, R., Page, N., Olivenstein, R., Elias, J., and Chakir, J. (2001). Il-17 is increased in asthmatic airways and induces human bronchial fibroblasts to produce cytokines. *J. Allergy Clin. Immunol.* **108**, 430–8.

61. Antonysamy, M. A., Fanslow, W. C., Fu, F., Li, W., Qian, S., Troutt, A. B., and Thomson, A. W. (1999). Evidence for a role of Il-17 in organ allograft rejection: Il-17 promotes the functional differentiation of dendritic cell progenitors. *J. Immunol.* **162**, 577–84.

62. Wu, C. Y., Kirman, J. R., Rotte, M. J., Davey, D. F., Perfetto, S. P., Rhee, E. G., Freidag, B. L., Hill, B. J., Douek, D. C., and Seder, R. A. (2002). Distinct lineages of T(H)1 cells have differential capacities for memory cell generation in vivo. *Nat. Immunol.* **3**, 852–8.

63. Powrie, F., Leach, M. W., Mauze, S., Menon, S., Caddle, L. B., and Coffman, R. L. (1994). Inhibition of Th1 responses prevents inflammatory bowel disease in Scid mice reconstituted with Cd45rbhi Cd4$^+$ T cells. *Immunity* **1**, 553–62.

64. Ito, H. and Fathman, C. G. (1997). Cd45rbhigh Cd4$^+$ T cells from Ifn-gamma knockout mice do not induce wasting disease. *J. Autoimmun.* **10**, 455–9.

65. Bregenholt, S., Brimnes, J., Nissen, M. H., and Claesson, M. H. (1999). In vitro activated Cd4$^+$ T cells from interferon-gamma (Ifn-gamma)-deficient mice induce intestinal inflammation in immunodeficient hosts. *Clin. Exp. Immunol.* **118**, 228–34.

66. Berg, D. J., Davidson, N., Kuhn, R., Muller, W., Menon, S., Holland, G., Thompson-Snipes, L., Leach, M. W., and Rennick, D. (1996). Enterocolitis and colon cancer in interleukin-10-deficient mice are associated with aberrant cytokine production and Cd4$^{(+)}$ Th1-like responses. *J. Clin. Invest.* **98**, 1010–20.

67. Pflanz, S., Timans, J. C., Cheung, J., Rosales, R., Kanzler, H., Gilbert, J., Hibbert, L., Churakova, T., Travis, M., Vaisberg, E., Blumenschein, W. M., Mattson, J. D., Wagner, J. L., To, W., Zurawski, S., McClanahan, T. K., Gorman, D. M., Bazan, J. F., de Waal Malefyt, R., Rennick, D., and Kastelein, R. A. (2002). Il-27, a heterodimeric cytokine composed of Ebi3 and P28 protein, induces proliferation of naive Cd4$^{(+)}$ T cells. *Immunity* **16**, 779–90.

68. Lucas, S., Ghilardi, N., Li, J., and De Sauvage, F. J. (2003). Il-27 regulates Il-12 responsiveness of naive Cd4$^+$ T cells through Stat1-dependent and -independent mechanisms. *Proc. Natl Acad. Sci. U S A* **100**, 15047–52.

69. Takeda, A., Hamano, S., Yamanaka, A., Hanada, T., Ishibashi, T., Mak, T. W., Yoshimura, A., and Yoshida, H. (2003). Cutting edge: role of Il-27/Wsx-1 signaling for induction of T-Bet through activation of Stat1 during initial Th1 commitment. *J. Immunol.* **170**, 4886–90.

70. Nieuwenhuis, E. E., Neurath, M. F., Corazza, N., Iijima, H., Trgovcich, J., Wirtz, S., Glickman, J., Bailey, D., Yoshida, M., Galle, P. R., Kronenberg, M., Birkenbach, M., and Blumberg, R. S. (2002). Disruption of T helper 2-immune responses in Epstein–Barr virus-induced gene 3-deficient mice. *Proc. Natl Acad. Sci. U S A* **99**, 16951–6.

71. Wang, J. and Fu, Y. X. (2005). Tumor necrosis factor family members and inflammatory bowel disease. *Immunol. Rev.* **204**, 144–55.

72. Atreya, R., Mudter, J., Finotto, S., Mullberg, J., Jostock, T., Wirtz, S., Schutz, M., Bartsch, B., Holtmann, M., Becker, C., Strand, D., Czaja, J., Schlaak, J. F., Lehr, H. A., Autschbach, F., Schurmann, G., Nishimoto, N., Yoshizaki, K., Ito, H., Kishimoto, T., Galle, P. R., Rose-John, S., and Neurath, M. F. (2000). Blockade of interleukin 6 trans signaling suppresses T-cell resistance against apoptosis in chronic intestinal inflammation: evidence in Crohn's disease and experimental colitis in vivo. *Nat. Med.* **6**, 583–8.

73. Strober, W., Fuss, I. J., and Blumberg, R. S. (2002). The immunology of mucosal models of inflammation. *Annu. Rev. Immunol.* **20**, 495–549.

74. Yamanaka, T., Helgeland, L., Farstad, I. N., Fukushima, H., Midtvedt, T., and Brandtzaeg, P. (2003). Microbial colonization drives lymphocyte accumulation and differentiation in the follicle-associated epithelium of Peyer's patches. *J. Immunol.* **170**, 816–22.

75. Onderdonk, A. B., Richardson, J. A., Hammer, R. E., and Taurog, J. D. (1998). Correlation of cecal microflora of Hla-B27 transgenic rats with inflammatory bowel disease. *Infect. Immun.* **66**, 6022–3.

76. Rath, H. C., Ikeda, J. S., Linde, H. J., Scholmerich, J., Wilson, K. H., and Sartor, R. B. (1999). Varying cecal bacterial loads influences colitis and gastritis in Hla-B27 transgenic rats. *Gastroenterology* **116**, 310–19.

77. Mizoguchi, A., Mizoguchi, E., Chiba, C., and Bhan, A. K. (1996). Role of appendix in the development of inflammatory bowel disease in Tcr-alpha mutant mice. *J. Exp. Med.* **184**, 707–15.

78. May, E., Lambert, C., Holtmeier, W., Hennemann, A., Zeitz, M., and Duchmann, R. (2002). Regional variation of the alphabeta T cell repertoire in the colon of healthy individuals and patients with Crohn's disease. *Hum. Immunol.* **63**, 467–80.

79. Kuckelkorn, U., Ruppert, T., Strehl, B., Jungblut, P. R., Zimny-Arndt, U., Lamer, S., Prinz, I., Drung, I., Kloetzel, P. M., Kaufmann, S. H., and Steinhoff, U. (2002). Link between organ-specific antigen processing by 20s proteasomes and Cd8$^{(+)}$ T cell-mediated autoimmunity. *J. Exp. Med.* **195**, 983–90.

80. Becker, C., Wirtz, S., Blessing, M., Pirhonen, J., Strand, D., Bechthold, O., Frick, J., Galle, P. R., Autenrieth, I., and Neurath, M. F. (2003). Constitutive P40 promoter activation and Il-23 production in the terminal ileum mediated by dendritic cells. *J. Clin. Invest.* **112**, 693–706.

Index